# Engineering Design with SolidWorks 2000

## A Competency Project Based Approach
## Utilizing 3D Solid Modeling

Marie P. Planchard & David C. Planchard

D0913754

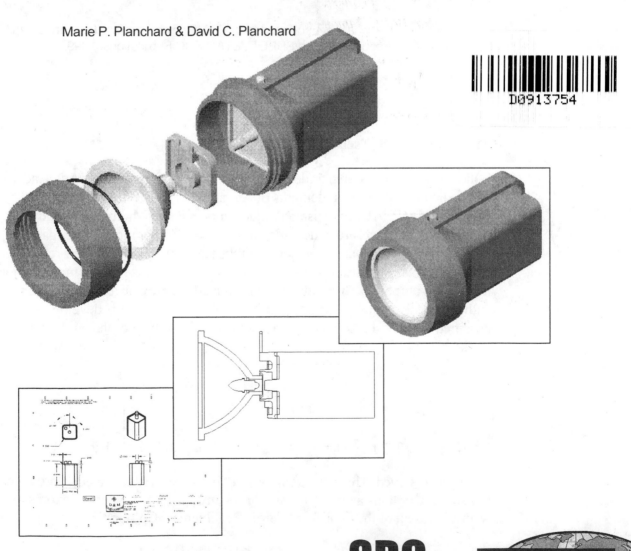

**SDC**
PUBLICATIONS

www.schroff.com

About the Authors:

Marie Planchard is Engineering Chair at Mass Bay in Wellesley Hills, MA. Before developing the CAD program, she spent 13 years in industry and held a variety of High Tech management positions including Beta Test Manager for CAD software at Computervision Corporation. She is Vice-President of the New England Pro/Users Group, an active member of the SolidWorks Educational Advisory Board, a SolidWorks Research Partner and coordinator for the New England SolidWorks Users Group.

David Planchard is the Director of Corporate Technology Programs at Middlesex Community College, in Bedford, MA. Before entering academia, he started his own Semiconductor Capital equipment company and spent 19 years in the Semiconductor industry in various Engineering and Marketing positions. He holds five U. S. Patents and one International. He is a member of the New England Pro/Users Group and the Cisco Regional Academy User group.

David and Marie Planchard are co-founders of D&M Engineering Inc. They are co-authors of the book, An Introduction to Pro/SHEETMETAL, SDC Publications, 1999.

**Trademarks and Disclaimer**

# INTRODUCTION

## Preface

Engineering Design in SolidWorks was written to assist students, designers, engineers and professionals. The book is focused on providing a solid foundation in SolidWorks using competency-based projects. Desired outcomes and usage competencies are listed for each project.

The book contains information for users to translate conceptual design ideas and sketches to manufacturing reality.

Commands are presented in a step by step progressive approach.

A competency project based curriculum is provided in the book. The learning process is explored through a series of design situations, industry scenarios, projects and objectives.

A progressive learning approach is addressed in each chapter. Each chapter identifies a project with reflective information on the previous project situation. The book compliments the on-line tutorials contained within SolidWorks.

## Notes to Instructors

Materials contained in this book can be covered in approximately 10 – 12 weeks, based upon a 4 credit, 15-week semester.  The remaining weeks are spent developing a final project.

Lecture time is focused on the development and review of manual sketches, conceptual layout sketches, industry scenarios, preliminary designs, modeling techniques and manufacturing concerns.

In the last three to five weeks of the semester, students should start their final project.  The project contains the following:

- Project plan

- Weekly status reports

- Concept sketches

- SolidWorks parts, assemblies, drawings and Bill of Materials

- Cost plan

- Manufacturing assembly procedure

- Engineering Change Order/Notice (ECO/ECN)

The book is also used in an accelerated 2 credit, 3D design class.

An Instructors guide will be available in the late summer of 2000.

## References:

References used in this text:

SolidWorks Users Guide, SolidWorks Corporation, 1999.

SolidWorks Tutorial, SolidWorks Corporations, 1999

ANSI Y14.5 Dimensioning and Tolerancing.

NBS Handbook 71, Specifications for Dry Cells and Batteries.

Betoline, Wiebe, Miller, Nasman, Fundamentals of Graphics Communication, Irwin, 1995.

Earle, James, Engineering Design Graphics, Addison Wesley, 1999.

Earle, James, Engineering Drafting, Creative Press, 1995.

Hoelscher, Springer, Dobrovolny, <u>Graphics for Engineers</u>, John Wiley, 1968.

Jensel & Helsel, <u>Engineering Drawing and Design</u>, Glencoe, 1990.

Ladouceur and McKeen, <u>Pro/E Solutions and Plastic Design</u>, Onward Press, 1999.

Lockhart & Johnson, <u>Engineering Design Communications</u>, Addison Wesley, 1999.

Meyer, Leo A., <u>Sheet Metal</u>, American Technical Publishers, Homewood, IL, 1995.

Olivo C., Payne, Olivo, T., <u>Basic Blueprint Reading and Sketching</u>, Delmar1988.

Planchard & Planchard, <u>An Introduction to Pro/SHEETMETAL</u>, SDC Publications, Mission, KS 1999.

Toogood, Roger, <u>Pro/ENGINEER Tutorial</u>, SDC Publications, Mission, KS 1999.

Walker, James, <u>Machining Fundamentals</u>, Goodheart Wilcox, 1999.

80/20 Product Manual, 80/20, Inc., Columbia City, IN, 1998.

GE Plastics Product Data Sheets, GE Plastics, Pittsfield, MA. 2000

Reid Tool Supply Product Manual, Reid Tool Supply Co., Muskegon, MI, 1999.

Simpson Strong Tie Product Manual, Simpson Strong Tie, CA, 1998.

## Acknowledgement

We would like to acknowledge the following individuals that have contributed guidance and advice to the design and content of this book. Their assistance has been invaluable.

- Stephen Schroff and Mary Schmidt, SDC Publications.
- Rosanne Kramer, SolidWorks Corporation.
- Sal Lama, SolidWorks Corporation.
- Dave Pancoast, SolidWorks Corporation.
- Fred Koehler, SolidWorks Corporation
- Robert McGill, SolidWorks Corporation
- Dana Seero, SolidWorks Reseller, Computer Aided Products, Inc.
- Keith Pedersen, Computer Aided Products, Inc.
- Jay Jacobs, Paperless Parts, Inc.

To the other professionals that provided their company information which was used in the book:

- Carl Kuo and Richard Chapman, Simpson Strong Tie.

- Russ MacIntyre, Lehi Sheetmetal.

- Jeff Pierce, Piece Aluminum.

- Tom Emery, Wilson Tool.

- Ron Clark, Reid Tool and Die.

- Ann Klapper

- Alan Barta, Brown and Sharpe.

- Robert Hess, GE Plastics.

- Bill Woomer, Air Inc.

- Dave Wood, 80/20 Inc.

- Gary Stahlinski ,GSC Engineering and Design Associates

- Richard Swenton, Sun Microsystems.

- Andy Roemer, LTX Corporation

Mass Bay engineering design students: Paul Bissaillon, BJ Klein, Rob Beaudette, Wendy Orton, Debrorah Kiziroglou, Richard Morrell, Carlos Bachiller, Ash Perkins, Greg Sylvia, Steve Lipkin, Todd Winner, Harold Jara, Thomas Leach, Jean Marc, Terence Mulhall, Sofia Savvidou, Kim Fontaine, Mario Bartoli, Allen Kessler and Joe Russo.

And Mass Bay CAD lab assistant, Allen Beaune for his editing assistance.

We value customer input. Please contact us with any comments, questions or concerns on this book.

Marie P. Planchard

Engineering Department Chair

Mass Bay Community College

planchar@mbcc.mass.edu

David C. Planchard

Director of Corporate Technology Programs

Middlesex Community College

planchardd@middlesex.cc.ma.us

# Dedication

To our loving and patient daughter. At the age of 8, just point her to the pencil shaped Sketch 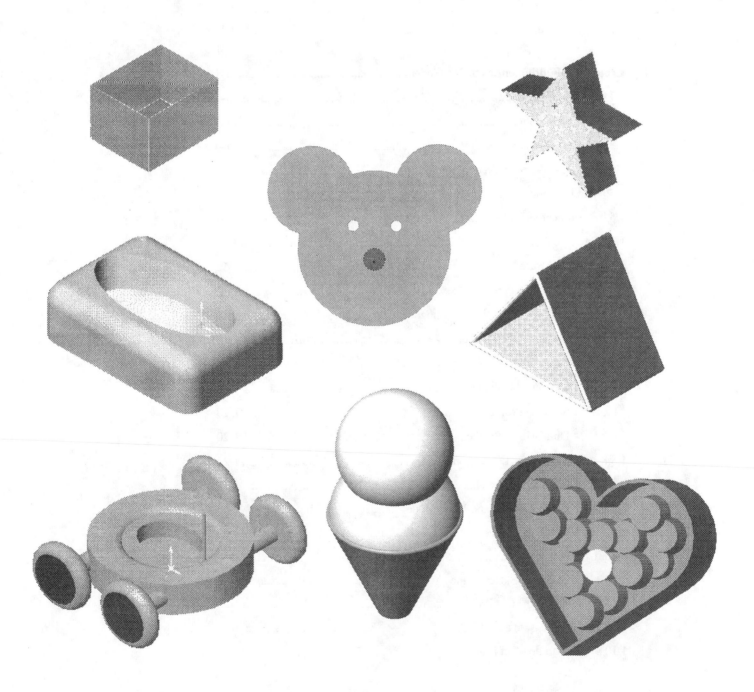 icon in SolidWorks and the fun begins.

Thank you Stephanie for the following illustrations.

## Table of Contents

## What is SolidWorks?

SolidWorks is a design automation software package used to produce parts, assemblies and drawings. SolidWorks is a Windows native 3D solid modeling CAD program. SolidWorks provides easy to use, highest quality design software for engineers and designers who create 3D models and 2D drawings ranging from individual parts to assemblies with thousands of parts.

SolidWorks Corporation, headquartered in Concord, Massachusetts, develops and markets innovative design solutions for the Microsoft Windows platform. More information on SolidWorks and its family of products can be found on their web site, www.SolidWorks.com.

In SolidWorks, you create 3D parts, assemblies and 2D drawings. The part, assembly and drawing documents are all related.

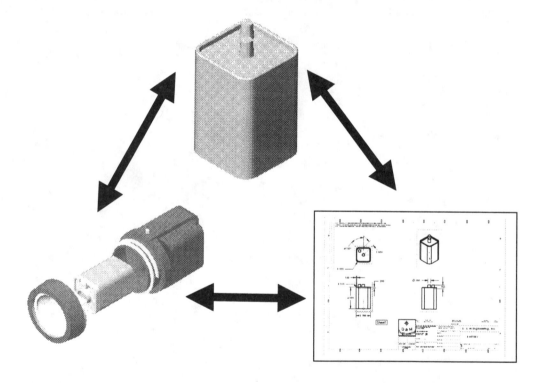

The building blocks of parts are called features. Features such as the *Extruded-Boss*, *Cut*, *Hole*, *Fillet* and *Chamfer* and others are used to create parts. Some features are sketched, such as an *Extruded-Boss*.

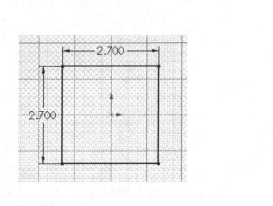 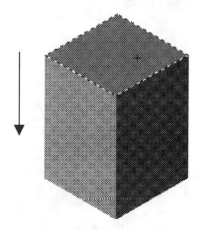

Other features are created by selecting edges or faces of existing features, such as a *Fillet*.

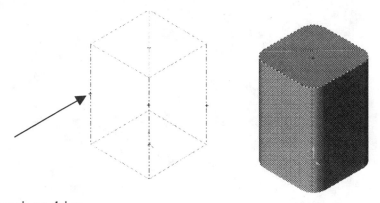

Dimensions drive features. Change a dimension and you change the size of the part.

Geometric relationships can be used to maintain the intent of the design.

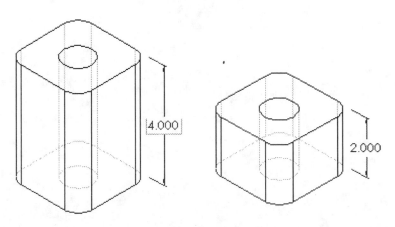

Create a hole that penetrates through the part. SolidWorks maintains the relationships through the change.

The step-by-step approach used in this text allows you to create parts, assemblies and drawings. The text allows you to modify and change all components of the model. Change is an integral part of design. The concept is introduced early in Project 1. Let's begin.

## Overview of Projects

Project 1 Fundamentals of 3D Solid Modeling in SolidWorks

How do you start a design in SolidWorks? What is the intent of the design? How do you take a customer's requirements and covert them into a model. Project 1 introduces the basic concepts behind SolidWorks. You create two parts: GUIDE and the ROD. You are exposed to the following features: *Extruded-Base*, *Extruded-Boss*, *Extruded-Cut*, *Fillet* and *Chamfer*.

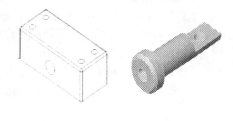

Project 2 Fundamentals of Assembly Modeling

Project 2 introduces the fundamentals of Assembly Modeling by creating the GUIDE-ROD assembly. You obtain a comprehensive understanding of incorporating design changes into an assembly. You create four FeaturePalette flange bolts to complete the assembly.

Project 3 Fundamentals of Drawing

Project 3 covers the development of a customized drawing template with your own logo. You create a GUIDE drawing with three standard views and an Isometric view. You develop and incorporate a Bill of Materials into the GUIDE-ROD drawing.

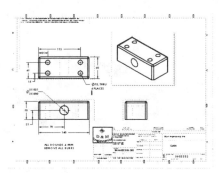

Project 4 *Extrude* and *Revolve* Features

Project 4 focuses on the customers design requirements to develop and FLASHLIGHT assembly. You create four key FLASHLIGHT components. The BATTERY and BATTERYPLATE are created with the *Base-Extrude* feature. The LENS and BULB are created with the *Base-Revolve* feature.

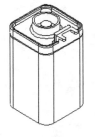

Project 5 Sweep and Loft Features

Project 5 continues the development of the
FLASHLIGHT assembly. You create four new
components. The O-RING utilizes a *Sweep*
feature. The SWITCH utilizes the *Loft* feature.
The LENSCAP and HOUSING apply features
presented in both Project 4 and Project 5.

Additional assembly techniques are developed
and the FLASHLIGHT assembly is created.

Project 6 Top Down Assembly

In Project 6 a Top Down approach is first
developed with a *Layout* Sketch. You create
components and modify them in the context of the
assembly. Sheet metal features are created to
develop a family of Electrical BOXES.

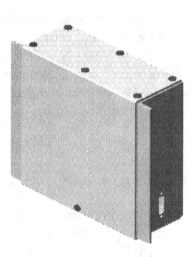

## Command Syntax

The following command syntax is used throughout the text. Commands that require you to perform an action are displayed in **Bold** text.

| Format | Convention | Example |
|---|---|---|
| **Bold** | All commands actions | Click **Save**.<br>Click **Tools**, **Options** from the Main menu. |
| | Selected icon button | Click the **Rectangle** ⊡ icon from the Sketch Tools toolbar. |
| | Selected geometry:<br>line, circle, arc, point<br>and text | Select the **centerpoint**.<br>Drag the **circle** downward.<br>Click the **arc**. |
| | Value entries | Enter **3.0** for Radius.<br>Click **60mm** from the Depth spin box.<br><br>Click **Blind** from the Type list box. |
| | Menu button entries | Click the **Horizontal** button. |
| | Filenames | Enter **GUIDE**. |
| | Key names | **Shift** key. **Ctrl** key |
| *Bold &*<br>*Italics* | Feature names | Click the ***Extruded-Base*** feature from the FeatureManager. |
| | Plane names | Click ***Top*** plane from the FeatureManager for the *Sketch* plane. |
| | View names | Click the ***Isometric*** view to insert a Bill of Materials |
| | Origin | Click the ***Origin***. |
| All CAPS | Filenames, part names,<br>assembly, component<br>and drawing names. | The BATTERY is contained inside the FLASHLIGHT assembly. |
| First Letter<br>Capitalized | Toolbar names | Sketch Tools toolbar. |
| | List box names | Click **Blind** from the Type list box. |
| | Menu names | Main menu. Pop-up menu. |
| **Bold Arial** | FeaturePalette names | Click **flange-bolt** from the FeaturePalette |
| Icon button | Square symbol that<br>represents a command | **Sketch** ✐ icon. |
| | | |

**NOTES:**

# Project 1

## Fundamentals of 3D Modeling

Below are the desired outcomes and usage competencies based upon completion
of Project 1.

| Project Desired Outcomes: | Usage Competencies: |
|---|---|
| • The understanding of the customer's design requirements and desires. | • To comprehend the fundamental definitions and process of Feature-Based 3D Solid Modeling. |
| • A product design that is cost effective, serviceable and flexible for future manufacturing revisions. | • Ability to translate initial design request and sketches into SolidWorks features: Base, Cut, Boss and Fillet/Round. |
| • Two parts:<br><br>    o ROD<br><br>    o GUIDE | • Ability to create and incorporate Edit features. |

## 1   Project 1 – Fundamentals of 3D Modeling

### 1.1   Project Objective

Create a GUIDE and ROD for a GUIDE-ROD assembly.

### 1.2   Project Situation

You receive a fax from the customer, Figure 1.0. The customer provides a concept drawing of a GUIDE-ROD assembly as a product request.

You review the fax. For Project 1, create the individual parts of the requested GUIDE-ROD assembly. Note: The assembly is a component used in a low volume manufacturing environment.

Figure 1.0

Before you begin, investigate a few key design questions:

1.   How will the customer use the assembly?

2.   How are the parts used in an assembly?

3.   Does this assembly affect other components?

4.   What are the design requirements for load, structural integrity or other engineering properties?

5.   What is the most cost effective material for each part?

6.   How will the parts be manufactured and what are their critical design features?

7.   How will each part behave when modified?

You may not have access to all of the required design information. Placed in a concurrent engineering situation, you are dependent upon others and are ultimately responsible for the final design.

Design information is provided from various sources. Ask questions. Part of the learning experience is to know which questions to ask.

The customer uses the ROD to position materials on a conveyor belt. The ROD is incorporated into a sub-assembly.

The ROD requires support. During the manufacturing operations, the ROD exhibits unwanted deflection. The engineering group calculates working loads on test samples. Material test samples include: 8086 Aluminum, 303 Stainless steel and a machinable plastic acetal. The engineering group recommends 303 Stainless steel for the ROD and GUIDE.

In the real world, there are numerous time constraints. The customer requires a quote, design sketches and a delivery schedule, YESTERDAY! If you wait for all of the required design information, you will miss the project deadline.

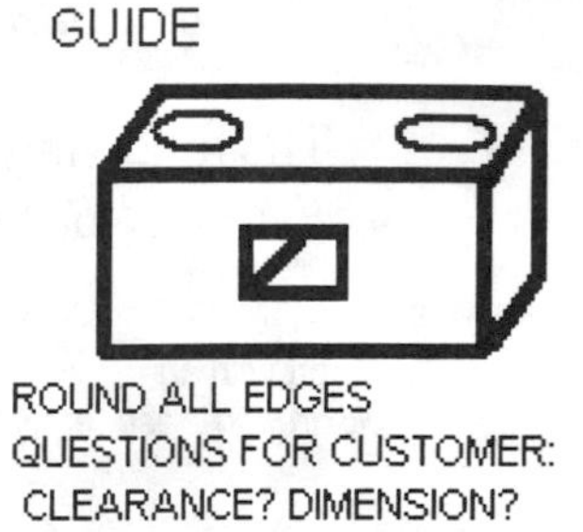

GUIDE

In a design review meeting, you create a rough concept sketch with idea notes, Figure 1.1a. Your colleagues review and comment on the concept sketch. Remember, this is a team environment.

ROUND ALL EDGES
QUESTIONS FOR CUSTOMER:
CLEARANCE? DIMENSION?

Figure 1.1a

In Project 1, the goal is to create the individual parts of the requested GUIDE-ROD assembly. In Project 2, the goal is to incorporate the individual parts into an assembly, Figure 1.1b.

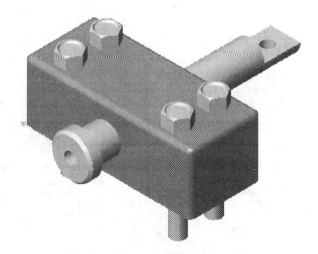

Figure 1.1b

### 1.3    Project Overview

Start the translation of the initial design sketch into SolidWorks features. Features are geometry building blocks.

Features add or remove material. Features are created from sketched profiles or from edges and faces of existing geometry.

The following features are used to create the GUIDE, Figure 1.2a: *Extruded-Base, Extruded-Cut* and *Fillet*.

The following features are used to create the ROD, Figure 1.2b: *Extruded-Base, Extruded-Cut, Chamfer and Extruded-Boss*.

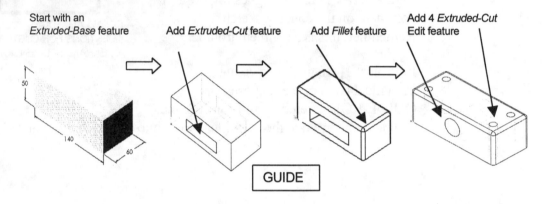

Figure 1.2a

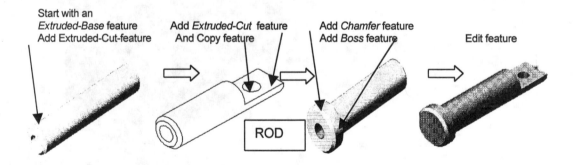

Figure 1.2b

Features can be modified, edited or deleted.

## 1.4    Windows 95/98/NT4.0 Terminology

The mouse pointer provides an integral role in executing SolidWorks commands. The mouse pointer executes commands, selects geometry, displays Pop-Up menus and provides information feedback.  A summary of mouse pointer terminology is displayed below:

| Item | Description |
|---|---|
| Click | • Press and release the left mouse button. |
| Double-click | • Double press and release the left mouse button. |
| Click inside | • Press the left mouse button. Wait a second and then press the left mouse button inside the text box.  This technique is used to modify Feature names in the FeatureManager design tree. |
| Drag | • Point to an object, press and hold down the left mouse button.  Move the mouse pointer to a new location, release the left mouse button. |
| Right-click | • Press and release the right mouse button.<br><br>• A Pop-up menu is displayed.  Use the left mouse button to select a menu command. |
| ToolTip | • Position the mouse pointer over an Icon (button).  The command is displayed below the mouse pointer. |
| Mouse pointer feedback | • Position the mouse pointer over various areas of the sketch: part, assembly or drawing.  The cursor provides feedback depending upon the geometry. |

Let's review various Windows terminology that describes: menus, toolbars and commands that constitute the graphical user interface in SolidWorks, Figure 1.3a, Figure 1.3b and Figure 1.3c.

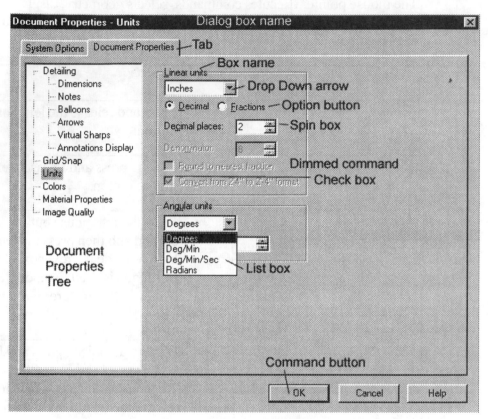

Figure 1-3a

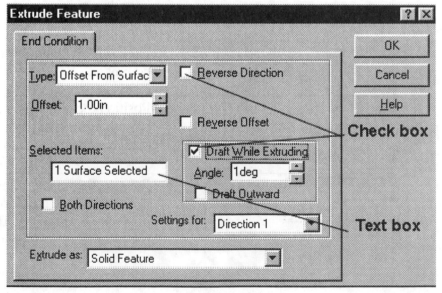

Figure 1-3b

| Item | Description of Windows Terminology |
|---|---|
| Dialog box name | • Name of a window to enter information in order to carry out a command. |
| Box name | • Name of a sub-window area inside the dialog box. |
| Check box | • Square box, click to turn on/off an option. |
| Spin box | • Box containing up/down arrows to scroll or type by numerical increments. |
| Dimmed command | • Menu command not currently available (light gray). |
| Tab | • Dialog box sub-headings to simplify complex menus. |
| Option button | • Small circle to activate / deactivate a single dialog box option. |
| List box | • Box containing a list of items. Click the list drop down arrow. Click the desired option. |
| Text box | • Box to type text. |
| Drop down arrow | • Opens a cascading list containing additional options. |
| OK | • Executes the command and closes the dialog box |
| CANCEL | • Closes the dialog box and leaves the original dialog box settings. |
| APPLY | • Executes the command. The dialog box remains open. |

Figure 1.3c

### 1.5    Part Creation

Create the GUIDE, Figure 1.4a.

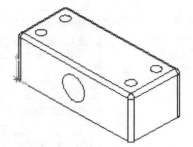

Click **Start** on the Windows Taskbar, .
Click **Programs**. Click the **SolidWorks 2000**
folder.

Click the **SolidWorks 2000** application.

**Figure 1.4a**

The SolidWorks program
window opens.

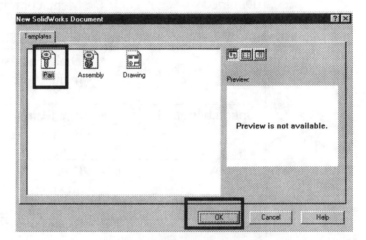

Create a part. Click the
**New** icon from the
Standard toolbar. Click
**Part** from the New
SolidWorks Document
dialog box, Figure 1.4b.
Click **OK**.

Part1 is displayed,
Figure 1.4c. Part1 is the
new default part window
name.

**Figure 1.4b**

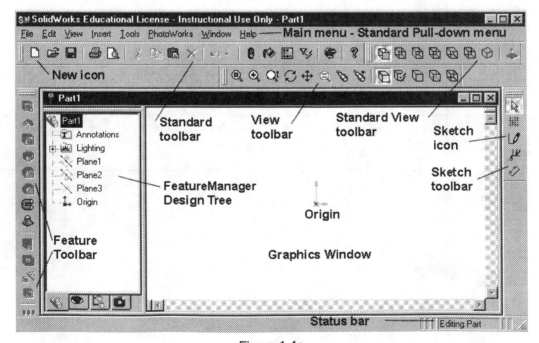

**Figure 1.4c**

Expand the SolidWorks window to full screen. Click the
**Maximize** icon in the top right hand corner of the Graphics
window.

Display the toolbars. Click **View** from the Main menu. Click **Toolbars**. The
system places a checkmark in front of the displayed toolbars.

Position the mouse pointer on an individual toolbar icon to receive a ToolTip.

Enter commands from the Toolbar icons, Main Pull down menu and or Short Cut
Keys. Examples:

- Toolbar icon                         Save ■ icon

- Main Pull down menu                  File, Save

- Short Cut Key                        Ctrl + S

For a summary of commands and additional information:

- Click the Help 🔲 icon

To simplify the commands in the book, the default
Toolbar icons and mouse Right-click Pop-up menus are
used throughout the text.

The Standard Main Pull down menu is referred to as the
Main menu throughout the remainder of the text.

## 1.6    Document and Grid/Snap Units

Document and Grid/Snap units assist the designer by
presenting a graphical drawing grid with selectable units
of measurement.

Set the document properties. Click **Tools** from the Main
menu. Click **Options**, Figure 1.5a.

Figure 1.5a

Set the document
units.  Click the
**Document Properties**
tab.  Click the **Units**
option.  The
Document
Properties – Units
dialog box is
displayed, Figure 1.5b.

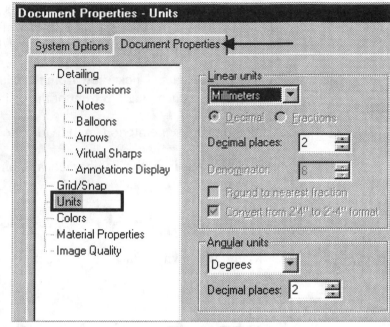

Click the **drop down
arrow** under the
Linear units box.
Select **Millimeters**
from the list box.
Enter **2** in the Decimal
places spin box.

Figure 1.5b

Set the grid units.
Click the **Grid/Snap**
option.  The Document Properties – Grid/Snap dialog box is displayed,
Figure 1.5c.  Click the **Display Grid** check box under the Grid Properties dialog
box.  Double-click the **value** in the Major grid spacing spin box.  Enter **100**.
Double-click the **value** in the Minor-lines spin box.  Enter **10**.

Click the **Snap
only when grid**
check box.
Click the **Snap
to points**
checkbox.

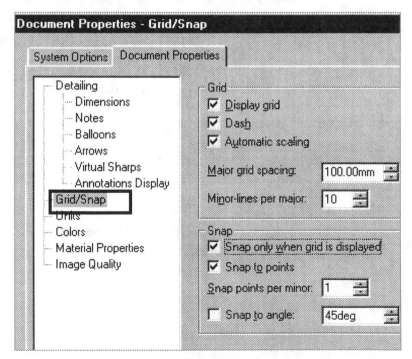

Figure 1.5c

Set the spin box increment. Click the **System Options** tab, Figure 1.5d. Click the **Spin Box Increment** option. The System Options – Spin Box Increments dialog box is displayed. Enter **5.00** inside the Metric units text box.

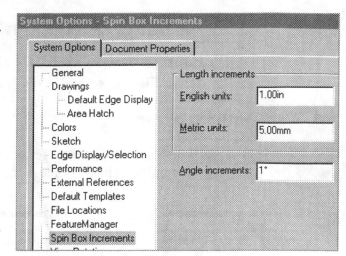

Figure 1.5d

Verify the General options. Click the **General** option, Figure 1.5e. Uncheck the **Input dimension value**. Check **Use shaded face highlighting**. Check **Enable Property Manager**.

All other System Options are unchecked. Update the current part document. Click **OK** from the System Options dialog box.

### 1.7  Save the Part

During a SolidWorks session, the first system default part filename is named: Part1.sldprt. The system attaches the .sldprt suffix to all created parts, Figure 1.6.

The second created part in the same session, increments to Part2.sldprt.

There are numerous ways to manage a part. Many companies use Part Data Management (PDM) systems to control, manage and document file names and drawing revisions.

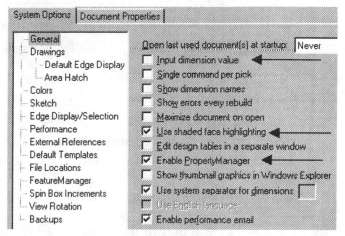

Figure 1.5e

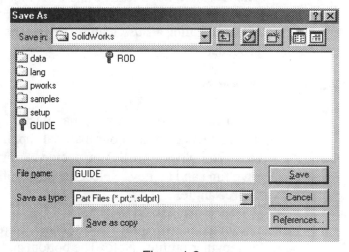

Figure 1.6

Save the part. Click the **Save** ■ icon. Enter the part name. Enter **GUIDE**. Click **Save**.

Use appropriate filenames that describe the part.

## 1.8   Extruded-Base Feature

What is a *Base* feature? The *Base* feature is the first feature that is created. The *Base* feature is the foundation of the part. Note: Keep the *Base* feature <u>simple!</u>

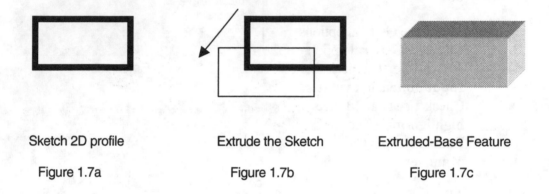

Sketch 2D profile          Extrude the Sketch          Extruded-Base Feature

Figure 1.7a               Figure 1.7b               Figure 1.7c

The *Base* feature geometry for the GUIDE is an extrusion. The extrusion is named *Extruded-Base* feature. To create a solid *Extruded-Base* feature:

- Sketch a rectangular profile on a flat 2D plane, Figure 1.7a

- Extend the profile perpendicular (⊥) to the *Sketch* plane, Figure 1.7b

The *Extruded-Base* feature is the 3D block, Figure 1.7c.

Before creating a *Base* feature, review the part manufacturing and assembly procedures.

### 1.8.1   Machined Part

In earlier conversations with manufacturing, a decision was made that the part would be machined. Your material supplier stocks raw material in rod, sheet, angle and block forms. You decide to start with a standard block form. This will save time and money. Select the best profile for the extrusion. This is a simple 2D rectangle.

Machined parts require datum planes for referenced dimensions. Sketch the rectangular profile. Note: The bottom and left edges are aligned to the reference planes.

Dimension holes and cuts from the same dimension reference planes, Figure 1.8a.

In 3D modeling, dimension the second hole from the edge of the *Extruded-Base* feature to its center point. Avoid dimension references between the center points of the holes, Figure 1.8b.

Referencing dimensions to the edge of the *Extruded-Base* feature provides information on how the part is manufactured and leads to fewer model rebuilding, "regeneration" problems and calculation errors.

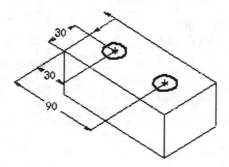 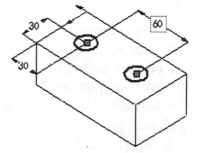

Figure 1.8a                              Figure 1.8b

### 1.8.2  Reference Planes and Orthographic Projection

The three default ⊥ reference planes, *Plane1, Plane2* and *Plane3* represent infinite 2D planes in 3D space, Figure 1.9. Planes have no thickness or mass.

Orthographic projection is the process of projecting views onto parallel planes with ⊥ projectors. The default reference planes are the *Front*, *Top* and *Right* viewing planes.

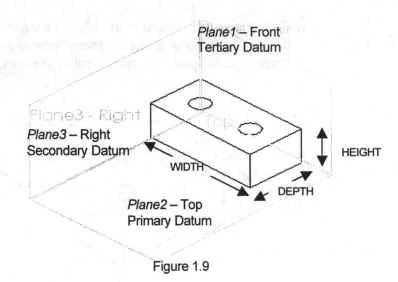

Figure 1.9

In geometric tolerancing, the default reference planes are the Primary, Secondary and Tertiary ⊥ datum planes, Figure 1.9. These are the same planes used in manufacturing.

- The Primary datum plane contacts the part at three or more points.

- The Secondary datum plane contacts the part at two or more points.

- The Tertiary datum plane contacts the part at one or more points.

Rename the default reference planes. Click *Plane1* in the FeatureManager design tree, Figure 1.10.

Click inside the *Plane1* text box. Edit the text. Enter *Front* in the text box. Enter *Top* for *Plane2*. Enter *Right* for *Plane3*.

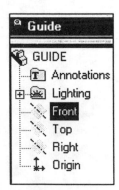

Figure 1.10

### 1.8.3   Create a Part Document Template

A part document template contains user-defined parameters such as:

- Reference plane names

- Grid units

- Part geometry

Use a part template to conserve setup time.

Create the part document template. Click **File** from the Main menu.  Click **Save As**.  Click **\*.prtdot** from the Save As type list box.  The default Templates file folder is displayed.  Enter

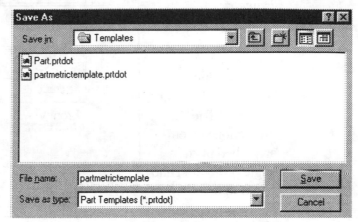

Figure 1-11a

**PARTMETRICTEMPLATE** in the File name text box, Figure 1.11a.  Click **Save**.

Verify the new template.  Click the **New** ▢ icon.  The new template, PARTMETRICTEMPLATE is added to the New SolidWorks Document Template dialog box, Figure 1-11b.

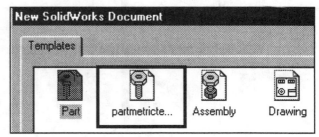

Figure 1-11b

Return to the GUIDE part.
Click **Cancel** from the New SolidWorks Document dialog box.

Note: Use the PARTMETRICTEMPLATE template when creating other metric parts in Project 1.

### 1.8.4  Orthographic Projection

In third angle Orthographic projection, the standard drawing views are *Front*, *Top*, *Right* and *Isometric*, Figure 1.12.

There are two Orthographic projection drawing systems. The projection systems are called third angle projection and first angle

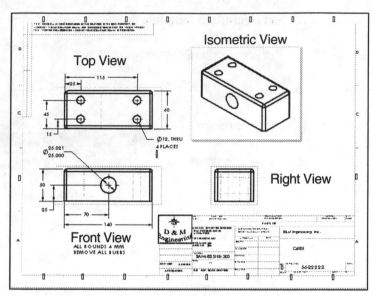

Figure 1.12

projection respectively, Figure 1.13. The systems are derived from positioning a 3D object in the third or first quadrant.

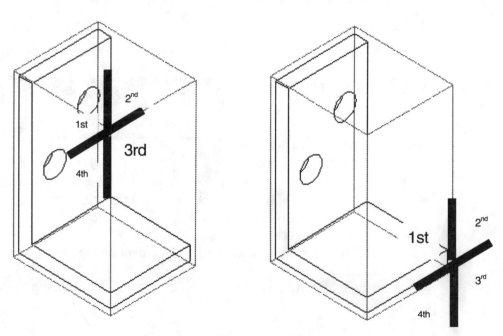

Figure 1.13

In third angle projection, the part is positioned in the third quadrant. The 2D projection planes are located between the viewer and the part. The projected views are placed on a drawing, Figure 1.14.

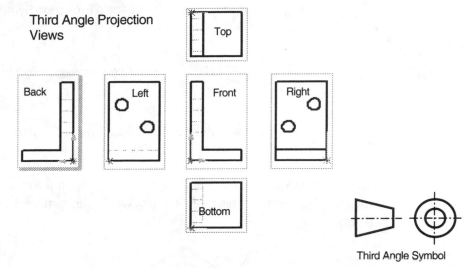

Figure 1.14

In first angle projection, the part is positioned in the first quadrant. Views are projected onto the planes located behind the part. The projected views are placed on a drawing, Figure 1.15.

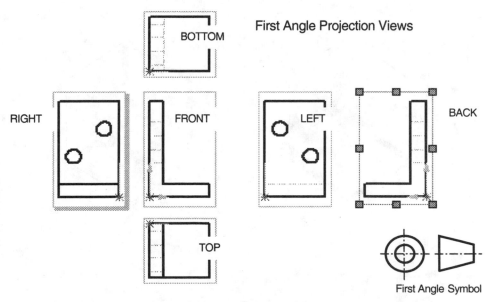

Figure 1.15

First angle projection is primarily used in Europe and Asia. Third angle projection is primarily used in the US and is the ANSI standard. Designers should have knowledge and understanding of both systems. There are many multi national companies. Example: A part is designed in the US, manufactured in Japan and destined for a European market.

Note: Third angle projection is used in this text. A truncated cone symbol appears on the drawing to indicate the projection system:

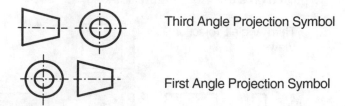

Third Angle Projection Symbol

First Angle Projection Symbol

The selected *Sketch* plane defines the orientation of the part in 3D space, Figure 1.16.

Create three examples by selecting the *Front*, *Top* and *Right* planes as *Sketch* planes. The overall dimensions are physically the same, however the orientation differs.

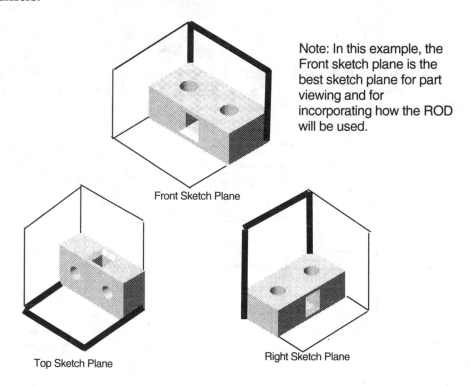

Note: In this example, the Front sketch plane is the best sketch plane for part viewing and for incorporating how the ROD will be used.

Front Sketch Plane

Top Sketch Plane

Right Sketch Plane

Figure 1.16

Before incorporating your design ideas into the *Sketch* plane, ask yourself a question:

• How will the part be oriented in the assembly?

A part should be oriented or aligned correctly to assist in the final assembly.

### 1.8.5  Create the Sketch

Creating a *Sketch* requires a *Sketch* plane and a 2D profile. Create a *Sketch* plane. Click the **Front** plane from the FeatureManager design tree, Figure 1.17a.

Create a new *Sketch*. Click the **Sketch** icon from the Sketch toolbar, Figure 1.17b.

Figure 1.17b

Figure 1.17a

The *Sketch* opens on the *Front* plane. The *Origin* represents the intersection of the *Front*, *Top* and *Right* planes.

The sketch grid displays the previously entered values in the Grid/Units dialog box, Figure 1.18.

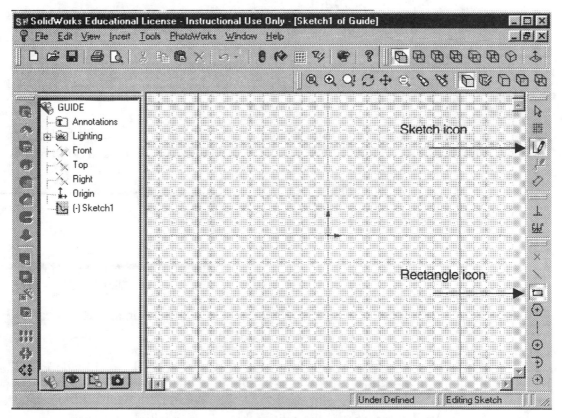

Figure 1.18

### 1.8.6   Sketch a Profile

Sketch the rectangle profile.  Click the **Rectangle** 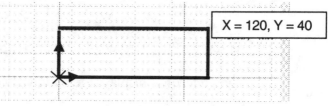 icon from the Sketch Tools toolbar.

The mouse pointer is displayed as a pencil with a rectangle ✏.

Note: When the mouse pointer is located on the *Origin,* the pointer displays a pencil with a point, ✏.

Create the first corner point.  Click the *Origin*.  Hold the left mouse button **down**. Create the second corner point.  Drag the **mouse pointer** up and to the right. Release the **left mouse button**.  The X-Y coordinates, (120,40) of the rectangle are displayed above the mouse pointer, Figure 1.19.

X = 120, Y = 40

Figure 1.19

Display the current Grid settings.  Click the **Grid** ▦ icon from the Sketch menu.  The current Grid/Snap properties are displayed.  Return to the Sketch.  Click **OK**.

Note: The Grid is turned off in the following illustrations for clarity.

### 1.8.7   Dimension a Profile

Dimensions add location and size information.  All models require dimensions for manufacturing.  Dimensions are not required to create features in SolidWorks.

Color and cursor feedback is used in SolidWorks to aid in the sketching process.  The current rectangular *Sketch* is displayed in two colors: blue and black, Figure 1.20.

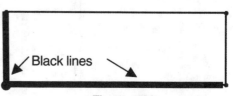

Black lines

Figure 1.20

The geometry consists of four lines and four vertices.  The vertex *Origin* is black.  The horizontal and vertical line positioned along the axes of the *Origin* is black.

In a fully defined *Sketch*, all entities are displayed in black.  A fully defined *Sketch* has a defined position, dimensions and or relationships.

In a under defined *Sketch*, the entities that require position, dimensions or relationships are displayed in blue.

Note: Click and drag under defined geometry to modify the *Sketch*.

In an over defined *Sketch*, there is geometry conflict with the dimensions and or relationships.

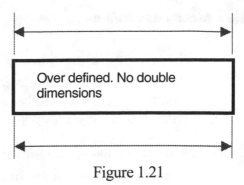

Figure 1.21

In an over defined *Sketch*, entities are displayed in red, Figure 1.21.

Dimension the horizontal line.

Click the **Dimension**  icon from the Sketch toolbar. The mouse pointer changes shape to the dimension symbol,  .

Note: The dimension text is modified in the next section.

Click the bottom **horizontal line** of the rectangle, Figure 1.22a. Drag the **mouse pointer** downward. Click the **dimension text** location below the black line.

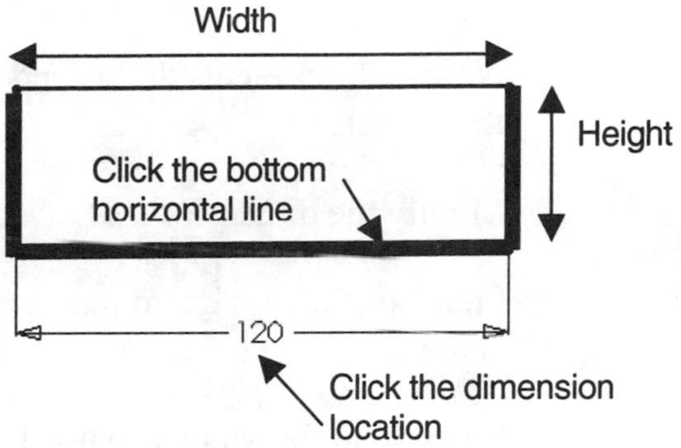

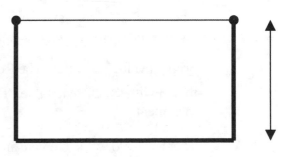

Figure 1.22a

The right vertical line is displayed in black. The bottom right vertex changes from blue to black. The width is fully defined. The top blue line that defines the height is under defined.

Click the **Select**  icon from the Sketch toolbar. Click the top **blue line**, Figure 1.23. Drag the **mouse pointer** upward. The vertical lines increase in size.

Figure 1.23

Dimension the horizontal line. Click the **Dimension**  icon from the Sketch toolbar. Click the **left most vertical line**. Drag the **mouse pointer** to the left. Click the **dimension text** location to the left of the *Sketch*. The *Sketch* is fully defined. All lines and vertices are displayed in black, Figure 1.24.

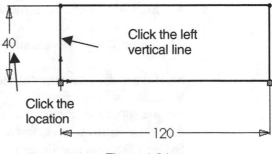

Figure 1.24

### 1.8.8  Modify the Dimension Values

Modify the dimension values to increase or decrease the size of the *Sketch*. Modify a dimension value. Click the **Select** icon from the Sketch toolbar or Right-click in the Graphics window and click Select, Figure 1.25a.

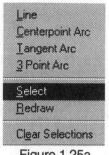

Figure 1.25a

Position the mouse pointer over the dimension text. The pointer changes to a linear dimension symbol, with a displayed text box D2@Sketch1 , Figure 1.25b.

Note: D2 represents the second linear dimension created in *Sketch*1.

Modify the dimension text. Double-click **40**. Click the Spin Box Arrows to increase or decrease dimensional values.

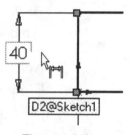

Figure 1.25b

Enter **45** in the Modify dialog box, Figure 1.26. Accept the value. Click the **Check** icon.

Figure 1.26

The Check icon saves the current value and exits the modify dialog box. The Restore icon restores the original value and exits. The Rebuild icon rebuilds the model with the current value. The Reset icon modifies the spin box increment.

There are two ways to create an exact dimension value. The first way is a two-step process:

- Step 1: Create the dimension

- Step 2: Modify the dimension

The second way is a one step process:

- Step 1: Use the Input dimension value option

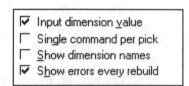

Figure 1.27

Set the Input dimension value. Click **Tools** from the Main menu. Click **Options**. Click the **Input dimension value** check box from the General option in the System Options tab, Figure 1.27. Return to the *Sketch*. Click **OK**.

The Modify dialog box appears automatically when the Dimension 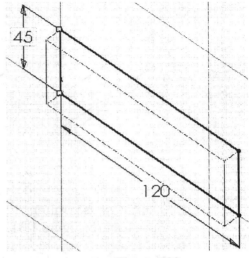 icon is selected with the *Sketch*.

The Input dimension value option saves time when creating many dimensions. This option is used for all projects in this text.

### 1.8.9    Extrude the Sketch

Extruding the *Sketch* adds depth to the rectangle.

Extrude the *Sketch*.  Click the **Extruded Boss/Base** icon in the Features toolbar. The *Extrude* feature dialog box is displayed.  The extruded *Sketch* is previewed in an *Isometric* view, Figure 1.28.

Figure 1.28

The preview displays the direction of the feature creation.

Reverse the direction of the extruded depth.  Click the **Reverse Direction** check box.  Specify the depth of the extrusion.  Blind is the default Type option.  Enter **60** in the Depth spin box, Figure 1.29.

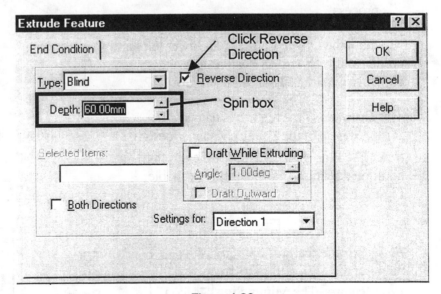

Figure 1.29

The depth of the extrusion increases respectively with the depth spin box value, Figure 1.30.

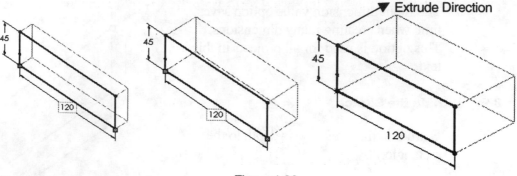

Figure 1.30

The *Origin* is positioned in the front lower left hand corner. Display the *Extruded-Base* feature, Figure 1.31. Click **OK**.

Note: Incorporate the machining and assembly process into the GUIDE design!

- Dimensions are referenced from the three datum planes with the machined *Origin* in the lower left hand corner of the block.

- Reference the front face of the GUIDE to the geometry on the ROD during the assembly process.

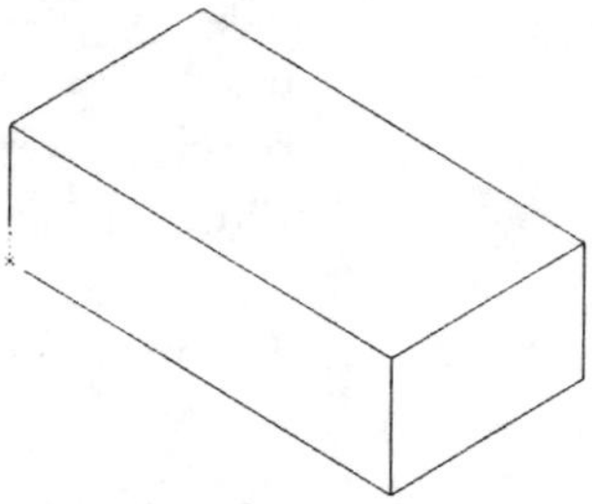

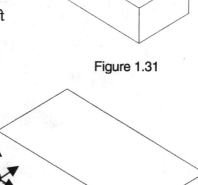

Figure 1.31

Maintain the *Origin* on the front lower left hand corner of the *Extruded-Base* feature and coplanar with the *Front* plane, Figure 1.32.

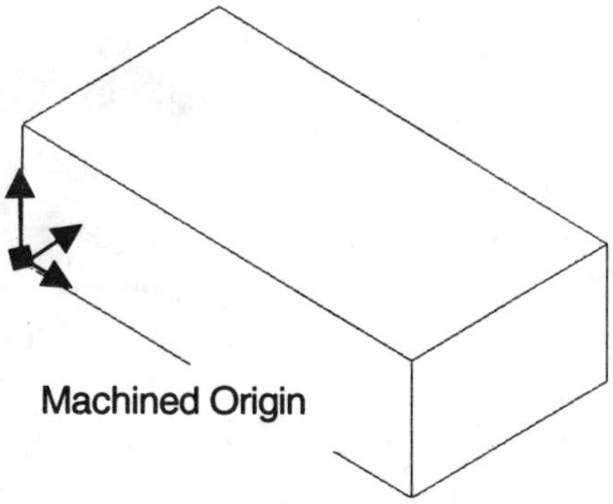

The *Extruded-Base* feature is named *Base-Extrude* in the FeatureManager design tree. The Plus Sign ⊞ icon indicates that additional feature information is available, Figure 1.33.

Machined Origin

Figure 1.32

Click the **Plus Sign** ⊞ icon of the *Base-Extrude* feature. *Sketch1* is the name of the *Sketch* used to create the *Base* feature. The Minus Sign ⊟ icon indicates that feature information is expanded. Click the **Minus Sign** ⊟ icon to collapse the *Base-Extrude* feature, Figure 1.34.

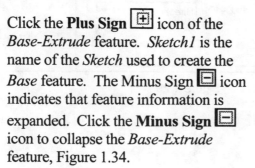

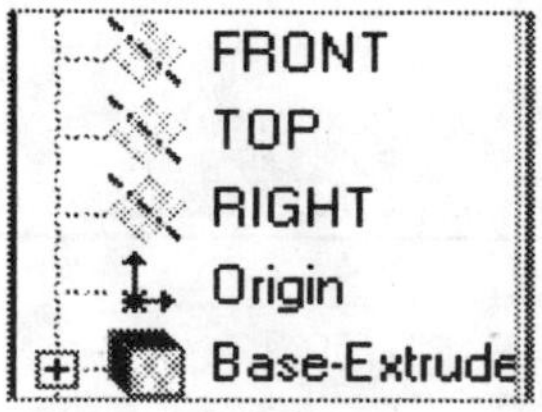

Figure 1.33

Figure 1.34

### 1.8.10  Display Modes and View Modes

Display modes create models in a variety of views, Figure 1.35.  Display the GUIDE.  Click the following icons from the View Display toolbar to assist in model visualization.

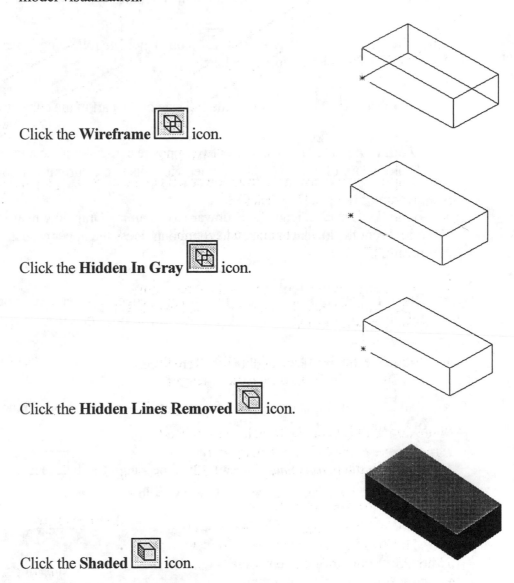

Click the **Wireframe** icon.

Click the **Hidden In Gray** icon.

Click the **Hidden Lines Removed** icon.

Click the **Shaded** icon.

Figure 1.35

Parts take longer to display as file size increases.  Use the Fast HLR/HLG icon to display models quickly.

The Fast HLR/HLG icon toggles between coarse and fine model display for the Hidden Lines Removed (HLR) and Hidden Lines Gray (HLG) mode.

Depending on your computer configuration and the part, the Hidden Lines Removed option will take longer to display than the Shaded option. The text examples use different display modes to illustrate features.

The view modes manipulate the model in the Graphics Windows.

Click the **Hidden Lines Removed** icon. Click the following icons from the View toolbar to assist in view visualization.

- **Zoom to Fit** icon displays the full size of the part in the current window.

- **Zoom to Area** icon displays two opposite corners of a rectangle to define the boundary of the view. The defined view fits to the current window.

- **Zoom In/Out** icon. Drag upward to zoom in. Drag downward to zoom out. Press the lower case **z** key to zoom out. Press the upper case **Z** key to zoom in.

- **Zoom to Selection** icon. Click the front edge. The selected geometry fills the current window. Select on a Vertex, an Edge or a feature.

- **Rotate** icon. Rotates about the screen center.

- **Pan** icon. Drag up, down, left, or right. The model scrolls in the direction of the mouse.

Return to the standard *Isometric* view. Click the **Zoom to Fit** icon. Click the **Isometric** icon from the Standards View toolbar.

Note: Right-click in the Graphics window area to display the zoom options, Figure 1.36.

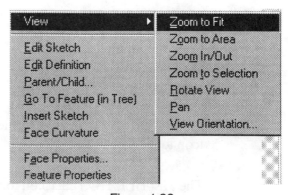

Figure 1.36

## 1.9   Modify Dimensions

Design changes and features are incorporated into the GUIDE.

Modify the GUIDE.

Double-click on the ***Base-Extrude*** feature.

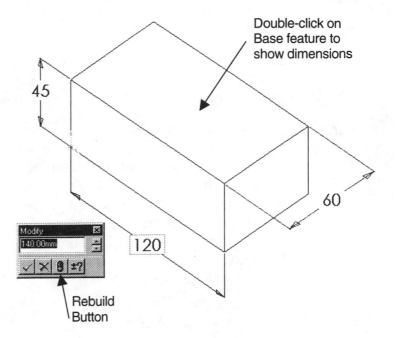

Figure 1.37a

Change the width dimension. Double-click on **120**. Enter **140** in the Modify dialog box, Figure 1.37a. Click the **Check mark** √ icon.

Change the height dimension. Double-click on **45**. Enter **50**.

Update the dimension. Click the **Rebuild** icon from the Modify dialog box, Figure 1.37b. Close the Modify dialog box. Click **Close**.

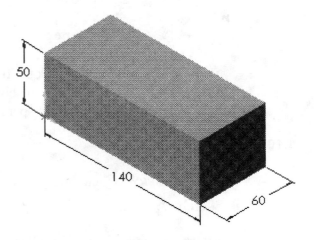

Figure 1.37b

Fit the model to the screen. Press the **f** key.

## 1.10  Cut Features

*Cut* features remove material from a part.  A *Cut-Extruded* feature requires a *Sketch* plane with one or more sketched profiles.  Note: A (2D) planar face or plane can be a *Sketch* plane.  Only sketch on one face or plane.

### 1.10.1  Select the Sketch Plane

Position the mouse pointer on the front face of the GUIDE, Figure 1.38.  Drag the mouse pointer over the part geometry.  The pointer displays various items depending upon geometry: Line, Face, Point or Vertex.  Example:

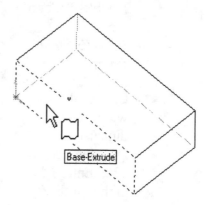

Figure 1.38

Select Line

Select Face

Select Point

Select Vertex

Select the *Sketch* plane.  Click the **front face** of the *Extruded-Base* feature.  When a face of a feature is selected, the boundary is highlighted with dash lines.  Note: Select the *Sketch* plane before opening a *Sketch*.  Sketch in the *Isometric* view.

### 1.10.2  Sketch the Cut Profile

Create a *Sketch*.  Click the **Sketch** 🖊 icon.

Sketch the rectangle profile.  Click the **Rectangle** ▱ icon from the Sketch Tools toolbar.  Click the **first point** in the lower left hand corner on the front GUIDE face, above the bottom edge of the *Extruded-Base* feature.

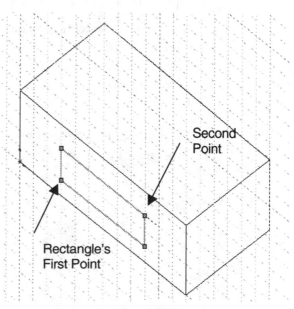

Second Point

Rectangle's First Point

Figure 1.39

Create the second corner point. Drag the **mouse pointer** up and to the right. Release the **left mouse button**, Figure 1.39.

### 1.10.3 Dimension a Profile

Various dimension schemes reflect the datum planes used in manufacturing.

- Dimension scheme 1: The overall cut dimensions reference an existing dimension, Figure 1.40a.

- Dimension scheme 2: Dimensions are referenced from the left side of the *Base* feature. The left side of the *Base* feature is co-linear with the *Right* plane, Figure 1.40b.

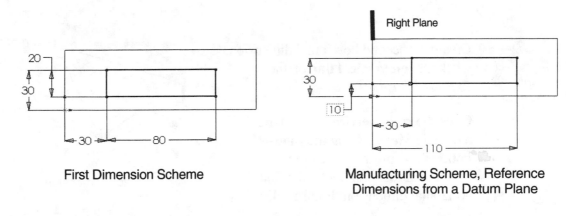

First Dimension Scheme          Manufacturing Scheme, Reference
                                          Dimensions from a Datum Plane

Figure 1.40a                                  Figure 1.40b

A critical feature of the GUIDE is the size of the *Cut*. The *Cut* feature is dependent on the ROD. To produce the cut, the machinist references all dimensions from the reference datum planes.

Note: Just as you reference all dimensions from the *Right* datum plane.

Display the *Right* plane, Figure 1.41. Right-click on the *Right* plane from the FeatureManager. Click **Show**.

The left side of the GUIDE is aligned to the *Right* plane. Hide the *Right* plane. Right-click on the ***Right*** plane from the FeatureManager. Click **Hide**.

Figure 1.41

Dimension the first horizontal dimension. Click the **Dimension** icon from the Sketch toolbar.

Select the first reference. Click the **left most vertical line** on the *Extruded-Base* feature.

Select the second reference. Click the rectangle *Sketch* left **vertical line**. Drag the **mouse pointer** downward. Click the **dimension** location below the *Extruded-Base* feature, Figure 1.42.

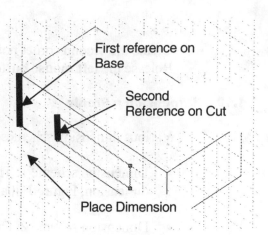

First reference on Base

Second Reference on Cut

Place Dimension

Figure 1.42

Create the second horizontal dimension. Click the **left vertical line** on the *Extruded-Base* feature.

Click the **right vertical line** on the rectangle *Sketch*. Drag the **mouse pointer** downward. Click the **dimension** location below the first horizontal dimension, Figure 1.43.

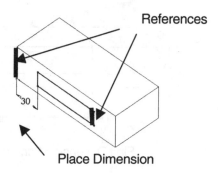

References

Place Dimension

Figure 1.43

Create the first vertical dimension. Click the bottom **horizontal line** of the *Extrude-Base* feature. Click the bottom **horizontal line** of the rectangle *Sketch*. Drag the **mouse pointer** toward the left. Click the **dimension** location to the left of the *Extruded-Base* feature, Figure 1.44.

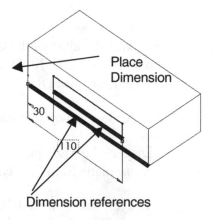

Place Dimension

Dimension references

Figure 1.44

Create the second vertical
dimension.  Click the **bottom
horizontal line** of the
*Extruded-Base* feature.  Click the
**top horizontal line** of the rectangle
*Sketch*.  Drag the **mouse pointer** to
the left.  Click the **dimension**
location to the left of the first
vertical dimension, Figure 1.45.

The *Sketch* is fully defined.  All
lines and vertices are displayed in
black!

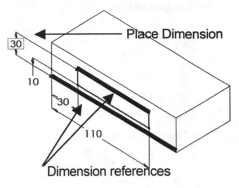

Figure 1.45

### 1.10.4 Extruded-Cut Feature

An *Extruded-Cut* feature removes material ⊥ to the *Sketch* for a specified depth.

Extrude the *Sketch*.  Click the **Extruded Cut** icon on the Features toolbar.  The
Extruded Cut feature dialog box displays and previews the extruded *Sketch*.  The
preview displays the direction of the feature creation.  The feature direction is into
the *Extruded-Base* feature.

Determine the depth.  Click the **drop down arrow** from the Type list box.  Click
the **Through All** option, Figure 1.46a.  Complete the *Extruded-Cut* feature.  Click
**OK**, Figure 1.46b.  View the *Extruded-Cut*.  Click the **Hidden In Gray** icon.

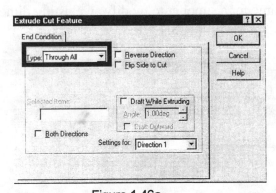

Figure 1.46a

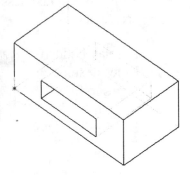

Figure 1.46b

The *Extruded-Cut* feature is named *Cut-Extrude1* in the FeatureManager.

Note: In the Through All option, the *Extruded-Cut* feature protrudes through
all ⊥ surfaces.

## 1.11 View Orientation

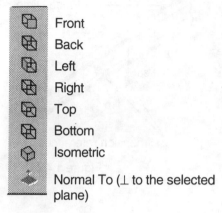

Front

Back

Left

Right

Top

Bottom

Isometric

Normal To (⊥ to the selected plane)

Figure 1.47

The view orientation defines the preset position of the model in the graphics window. The Standard View toolbar displays eight view options: *Front, Back, Left, Right, Top, Bottom, Isometric* and *Normal To*, Figure 1.47.

The *Isometric* view displays the part in 3D with two equal projection angles. The *Normal To* view displays the part ⊥ to the selected plane.

View orientation is selected for sketches, parts and assemblies.

## 1.12 Fillet/Round Feature

Sharp edges and square corners can create part stress. The *Fillet/Round* feature removes sharp edges, strengthens corners and/or cosmetically improves appearance. *Fillets* blend inside surfaces. *Rounds* blend outside surfaces.

On castings and plastic modeled parts, implement *Fillets* and *Rounds* into the initial design.

If you are uncertain of the exact radius value, input a small test radius of 0.05" or 1mm. It will take less time for a supplier to modify an existing dimension than to create one.

Display Hidden Lines to select edges to fillet. Click the **Hidden In Gray** icon.

Create a *Fillet/Round* feature. Click the **Fillet/Round** icon. The *Fillet* feature dialog box appears, Figure 1.48.

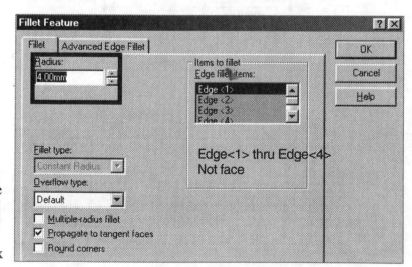

Figure 1.48

Enter **4** in the Fillet Radius list box.  Click the **4 vertical edges**, Figure 1.49.  Each edge is added to the Items to Fillet list.  Display the edge *Fillet*, Figure 1.50. Click **OK**.

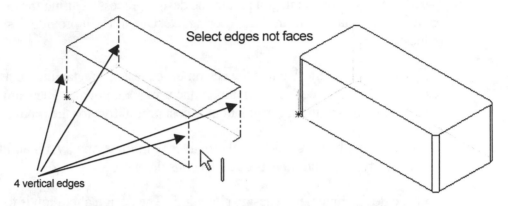

Select edges not faces

4 vertical edges

Figure 1.49                                        Figure 1.50

Minimize the number of features created in the FeatureManager. Combine *Fillets* and *Rounds* that have a common radius. Select all *Fillet/Round* edges.  Add them to the Items to Fillet list in the *Fillet/Round* feature dialog box.

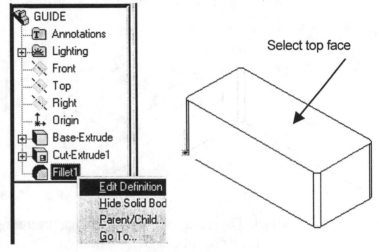

Select top face

In the previous step, you selected edges to *Fillet*.  Faces with the same fillet radius can be added to the Items to Fillet list using the Edit Definition option.

Figure 1.51

Add a new face to the *Fillets*.  Right-click the *Fillet* feature from the FeatureManager, Figure 1.51.  Click **Edit Definition**.  The Fillet Feature dialog box is displayed.  Click the **top face** of the *Extruded-Base* feature.  The face is added to the Items to Fillet.  Display the face *Fillet*, Figure 1.52.  Click **OK**.

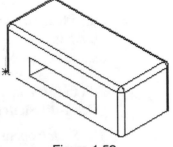

Figure 1.52

Modify feature size by changing dimensions.
Edit feature parameters with the Edit Definition command.

Save the GUIDE. Click the **Save** ⊟ icon.

## 1.13 Design Changes

Design changes are an integral part in the design process. During the design stage, the customer, manufacturing, engineering, etc. can and will provide design change requests.

The design team identified that the customer requested a rectangular cut. The rectangular cut will require special manufacturing equipment. The team decides that the expense for the rectangular cut is cost prohibitive for the product.

The team re-evaluates the design alternatives with functionality, reliability, serviceability, cost and time as essential considerations.

You decide to implement a design alteration. The design alteration is to change the rectangular cut to a circular cut, Figure 1.53. The advantage of the design alteration is that the hole requires no special manufacturing equipment.

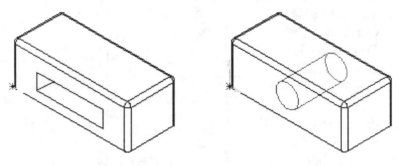

Figure 1.53

The design change will lower the manufacturing cost. The alteration does not effect any critical design specification. You propose the design change to the customer. The customer agrees!

## 1.14 Edit a Feature

Edit a feature by modifying its initial *Sketch*. The initial *Sketch* is the profile used to create the feature. The feature definition is defined by the feature parameters.

Figure 1.54

Edit the *Sketch*. Right-click the ***Cut-Extrude1*** feature. Click **Edit Sketch** from the Pop-up menu, Figure 1.54.

The *Sketch* opens. The Sketch ✏ icon is automatically selected. The rectangular *Sketch* is displayed in black.

Display the *Front* view orientation. Click the **Front view** 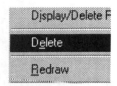 icon from the Standard View toolbar.

Delete the rectangular profile. Hold the **Ctrl** key down. Click the **4 lines**, Figure 1.55a. The lines are displayed in green.

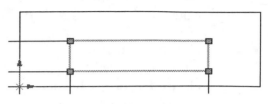

Figure 1.55a

Note: The Ctrl key allows you to select multiple objects at one time.

Delete the vertical lines. Right-click in the Graphics Window, Figure 1.55b. Click **Delete**. Release the **Ctrl** key.

Figure 1.55b

The system displays the Sketcher Confirm Delete dialog box, Figure 1.56. The vertical and horizontal dimensions for the rectangle reference the *Extruded-Base* feature. Delete the rectangle dimensions. Click **Yes to All**.

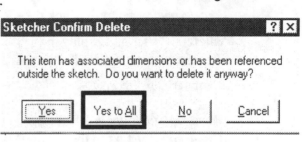

Figure 1.56

Sketch the circle profile.

Click the **Circle** icon from the Sketch Tools toolbar. Click the **center** of the front face of the *Extruded-Base* feature. Create the second point of the circle. Drag the **mouse pointer** to the right. Release the **left mouse button**, Figure 1.57.

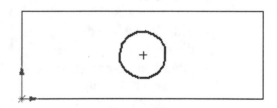

Figure 1.57

Create the horizontal linear dimension. Click the **Dimension** icon. Click the **left edge of the *Extruded-Base*** feature. Click the **center point** of the circle. Click the **text** location below the *Sketch*, Figure 1.58. Enter **70**.

Click the **Check** icon.

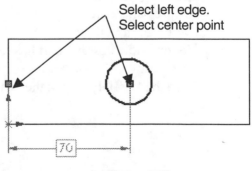

Figure 1.58

Create the vertical
dimension. Click the
**bottom edge**. Click the
**center point**. Click the
**text** location to the left of
the *Sketch*, Figure 1.59.
Enter **25**.

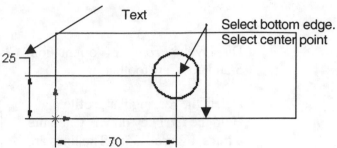

Figure 1.59

Create the diameter
dimension. Click the
**circumference**. Click the
**text** location below the *Sketch*,
Figure 1.60. Enter **25**.

The circle dimensions are
completely defined. The *Sketch*
is displayed in black.

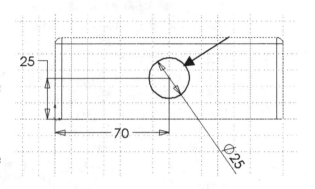

Display the updated
*Cut-Extrude1* feature. Click the
**Sketch** 🖋 icon. Display the
*Isometric* view. Click the
**Isometric** ◈ icon.

Figure 1.60

Rename *Cut-Extrude1* feature to describe the created geometry. Click the
***Cut-Extrude1*** text in the FeatureManager. Click inside the ***Cut-Extrude1*** text
box. Enter ***GuideHole***,
Figure 1.61.

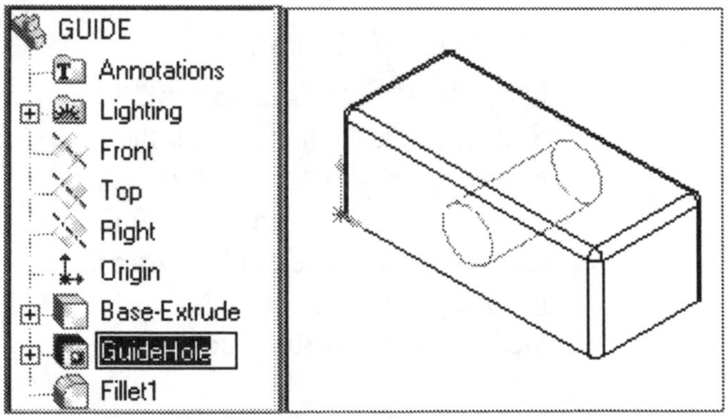

The GuideHole name is
displayed in blue. All
other feature names are
displayed in yellow.
Update the GUIDE.

Click the **Rebuild** 🔾 icon
from the Main menu. All
feature names are
displayed in yellow.
Yellow indicates that part has rebuilt correctly.

Figure 1.61

Save the GUIDE. Click the **Save** 💾 icon.

### 1.15 Fasteners

Screws, bolts and fasteners are used to joint parts together. Use standard available fasteners whenever possible. This will decrease product cost and reduce component purchase lead times. The American National Standard Institute (ANSI) and the International Standardization Organization (ISO) provide standards on various hardware components. The SolidWorks Library contains a variety of standard fasteners to choose from.

Below are general selection and design guidelines that are utilized in this text:

- Use standard industry fasteners where applicable.

- Reuse the same fastener types where applicable. Dissimilar screws and bolts require different tools for assembly, additional part numbers and increase inventory storage and cost.

- Decide on the fastener type before creating holes. Dissimilar fastener types require different geometry's.

- Create notes on all fasteners. This will assist you in the development of a Parts list and Bill of Materials.

- Caution should be used in positioning holes. Do not position holes too close to an edge. At a minimum, stay one radius head width from an edge or between holes.

- Design for service support. Insure that the model can be serviced in the field or on the production floor.

In this exercise, use standard M10 x 100 flat socket thread cap screws. M10 represents a metric screw - 10 mm major outside diameter, 100 mm overall length.

### 1.16 Holes as Cut Features

Holes remove material. There are many ways to create Holes. In this example, the *Cut* feature is used to create holes.

The GUIDE requires mounting holes. The holes are machined through the top surface.

Select the *Sketch* plane. Click the **top face** of the *Extruded-Base* feature. Dash lines are displayed around the face perimeter, Figure 1.62.

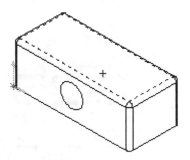

Figure 1.62

Create a *Sketch*. Click the **Sketch** icon.
Display the *Top* view, Figure 1.63. Click the
**Top** icon.

The *Origin* is located in the lower left corner.
Display the reference plane location with
respect to the *Sketch* plane. Click the ***Front***
and ***Right*** plane from the FeatureManager
design tree.

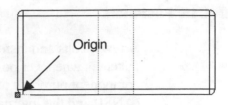

Figure 1.63

Create the first circle. Create Circle1. Click
the **Circle** icon from the Sketch Tools
toolbar. Identify the first circle point. Click
the **center point** on the right side of
*GuideHole*, Figure 1.64.

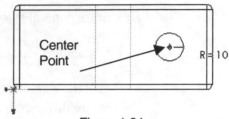

Figure 1.64

Create the second circle point. Drag the **mouse pointer** to the right, until R = 10.
Release the **mouse button**. The radius value is displayed in millimeters. The
circle is green.

Note: The display R=10 occurs when the grid snap to point is turned on and the
mouse is dragged horizontally.

### 1.16.1 Move and Copy Sketch Geometry

Move and copy sketch geometry to create four circles.

The circumference and center point are currently selected. Right-click in the
**Graphics window**. Click **Select** from the Pop-up menu, Figure 1.65. The center
point and circle circumference is blue.

Click the **center point** of the circle, Figure 1.66. The center point is green and the
circumference is blue.

Drag the **center point** of Circle1 upward. Release the **mouse button**, Figure 1.67.

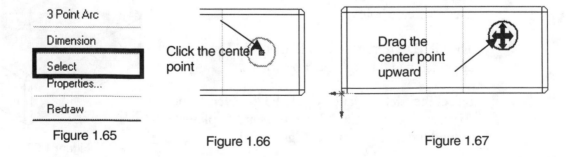

Figure 1.65                     Figure 1.66                     Figure 1.67

Use the Ctrl key to copy Circle1. Create Circle2. Hold down the **Ctrl** key. Click and drag the **circumference** of Circle1. Drag **Circle2** to the left. Release the **mouse button**. Release the **Ctrl** key, Figure 1.68.

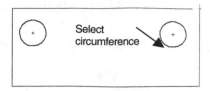

Figure 1.68

Copy Circle1 and Circle2. Hold the **Ctrl** key

down. Select the **circumference** of Circle1. Circle1 and Circle2 are displayed in green.

Click the **circumference** of Circle1. Drag the **mouse pointer** downward. Place Circle3 and Circle4, Figure 1.69. Release the **mouse button**.

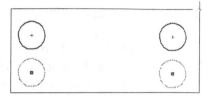

Figure 1.69

## 1.16.2 Geometric Relations

Geometric relations are dimensional relationships between one or more creations. The holes may appear aligned and equal, but they are not!

Click the **Geometric Relations** icon from the Add Geometric Relations toolbar.

The two bottom circles, Circle3 and Circle4 are selected system entites, named Arc3 and Arc4.

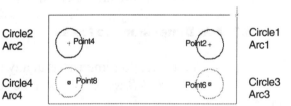

Circle2   Point4       Point2   Circle1
Arc2                             Arc1

Circle4   Point8       Point6   Circle3
Arc4                             Arc3

Figure 1.70

Note: The SolidWorks default name for curve geometry is an Arc#. There is an Arc# for each circle.

Click the **Circle1** circumference. Click the **Circle2** circumference, Figure 1.70. The system default name for these circles are Arc1 and Arc2 respectively. Click the **Equal** button. This maintains the same diameter for all four circles. Click **Apply**, Figure 1.71.

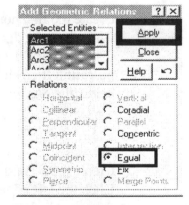

Figure 1.71

Note: If you selected the center point and not the circumference, right-click in the Graphics window and click Clear Selections. All geometry is removed from the Selected Entites text box. Click only the circumference of the four circles.

If you create or delete the geometry in a different order, the system selected entity names will be different.

The two bottom center points are system entites named Point6 and Point8. The default name for any point is an Point#. Each circle requires two points: a center point and a point on the diameter.

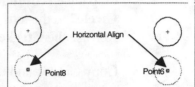

Align the center points. Click the two bottom center points: **Point6** and **Point8**, Figure 1.72a. Click the **Horizontal** button for horizontal point alignment. Click **Apply**, Figure 1.72b.

Figure 1.72a

Click the two top center points: **Point2** and **Point4**. Click the **Horizontal** button for horizontal point alignment. Click **Apply**.

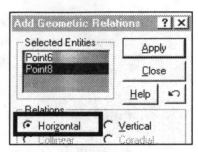

Click the two left center points: **Point4** and **Point8**. Click the **Vertical** button for vertical point alignment. Click **Apply**.

Click the two right center points: **Point2** and **Point6**. Click the **Vertical** button for vertical point alignment. Click **Apply**. Click **Close**.

Figure 1.72b

### 1.16.3 Add Dimensions to Holes

You can add dimensions at any time. Each dimension references the edge of the *Extruded-Base* feature and the center point of the hole.

Dimension the holes. Click the **Dimension** ⟋ icon.

Create the first horizontal dimension. Click the **left edge of the *Extruded-Base*** feature. Click the **center point of the bottom left circle**. Click the **text** location below the *Sketch*, Figure 1.73a. Enter **15**.

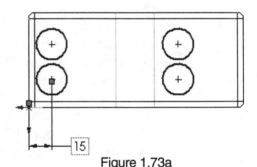

Figure 1.73a

Create the second horizontal dimension. Click the **left edge of the *Extruded-Base*** feature. Click the **center point of the bottom right circle**. Click the **text** location below the *Sketch*, Figure 1.73b. Enter **125**.

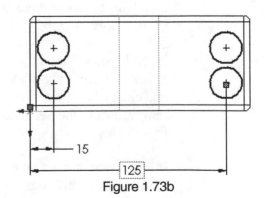

Figure 1.73b

Note: If the *Sketch* is too large in the Graphics window you cannot see the dimension. Press the f key to fit the *Sketch* to the screen.

Create a vertical dimension. Click the **bottom edge of the** *Extruded-Base* feature. Click the **center point of the bottom left circle**. Click the **text** location left of the *Sketch*, Figure 1.73c. Enter **15**.

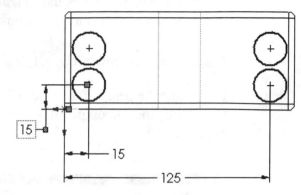

Figure 1.73c

Create a vertical dimension. Click the **bottom edge of the** *Extruded-Base* feature. Click the **center point of the top left circle**. Click the **text** location left of the *Sketch*, Figure 1.73d. Enter **45**.

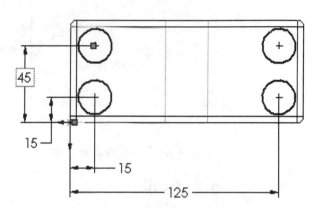

Figure 1.73d

Create a diameter dimension. Click the **circumference of the top right circle**. Drag the **dimension** above the *Sketch*. Enter **10**. All 4 holes are modified from ∅20 to ∅10. Note: Their geometric relationship was set to Equal, Figure 1.74.

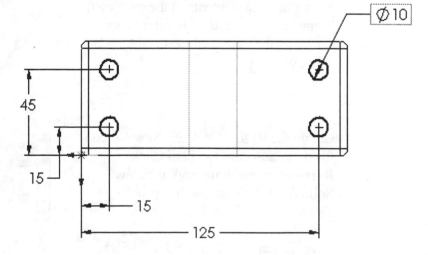

Figure 1.74

### 1.16.4 Extrude the Sketch

Extrude the *Sketch*. Click the **Extruded-Cut**  icon on the Features toolbar. The *Extruded-Cut* feature dialog box is displayed, Figure 1.75a.

Display the *Isometric* view. Click the **Isometric** icon.

Determine the depth of the hole. Click the **drop down arrow** from the Type list box. Click the **Through All** option. Create the four holes. Click **OK**, Figure 1.75b.

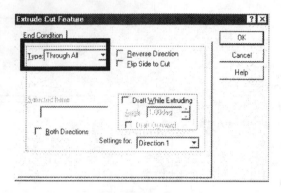

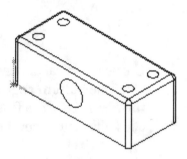

Figure 1.75a                                    Figure 1.75b

Rename *Cut-Extrude2* to *Holes*.

**Save** the GUIDE.

## 1.17 ROD Creation

Recall the requirements of the customer, Figure 1.76. The ROD is part of a sub-assembly that positions materials onto a conveyor belt.

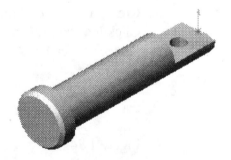

Figure 1.76

Create the ROD. Click the **New** icon from the Standard toolbar. Click **Partmetrictemplate** from the New SolidWorks Document dialog box, Figure 1.77a. Click **OK**.

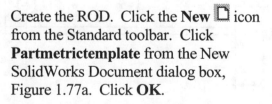

Figure 1-77a

Save the part. Click the **Save** icon. Enter the part name. Enter **ROD**. Click **Save**.

The FeatureManager displays the part name, ROD, Figure 1.77b.

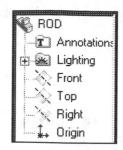

Figure 1.77b

The reference names, *Front*, *Top* and *Right* and the document grid/units settings are stored with the partmetric template.

Input the dimension value at creation. Click **Tools** from the Main menu. Click **Options**. Click the **Input dimension value** check box from the General System Option. Click **OK**.

### 1.17.1 Extruded-Base Feature

The geometry of the *Base* feature is a cylindrical extrusion, Figure 1.78. The *Extruded-Base* feature is the foundation for the ROD.

What is the shape of the sketched 2D profile? Answer: A circle. What is the *Sketch* plane? Before you answer the question, remember how the ROD is positioned in the assembly. Answer: The *Front* plane is the *Sketch* plane.

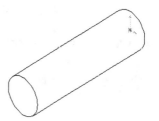

Figure 1.78

### 1.17.2 Create the Sketch

Select the *Sketch* plane. The *Front* plane is the default

*Sketch* plane. Open the *Sketch*. Click the **Sketch** icon from the Sketch toolbar.

Create a circle. Click the **Circle** icon from the Sketch Tools toolbar. Create the first point. Click the

*Origin* . Create the second point. Drag the **mouse pointer** to the right until the value displayed on the pointer is R = 20. Release the **mouse button**. Click the

**Dimension** icon. Click the **circumference** of the circle. Enter **25**, Figure 1.79.

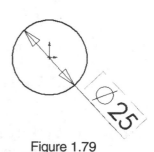

Figure 1.79

### 1.17.3 Create the Extruded-Base Feature

Extrude the *Sketch*. Click the

**Extruded Boss/Base** icon on the Features toolbar. The Extrude Feature dialog box is displayed, Figure 1.80.

Blind is the default Type option. Enter **120** in the depth text box.

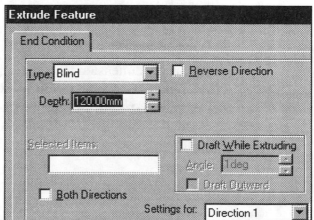

Figure 1.80

Display the *Extruded-Base* feature, Figure 1.81. Click **OK**.

View the *Extruded-Base* feature. Click the **Zoom to Fit** 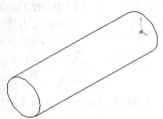 icon.

Figure 1.81

## 1.18 Extruded-Cut Feature

Create a hole. Select the *Sketch* plane. Click the **front circular face** of the ROD, Figure 1.82.

Click the **Normal To** icon. The front face of the ROD is displayed, Figure 1.83.

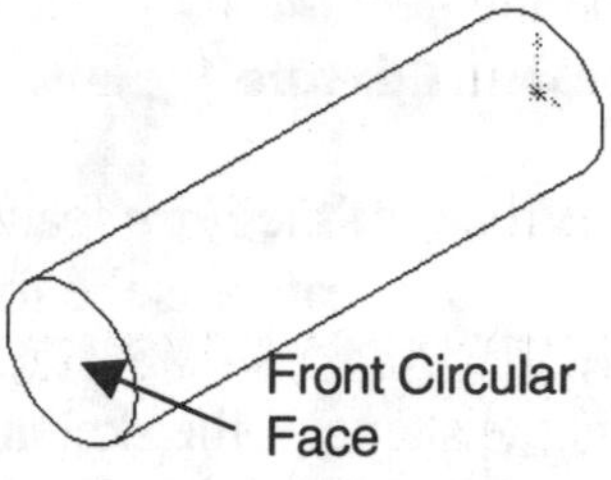

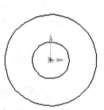

Front Circular Face

Figure 1.82 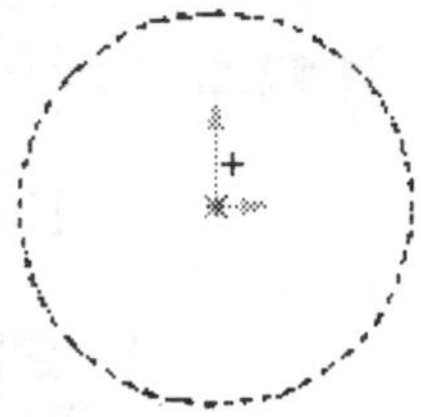          Figure 1.83

## 1.18.1 Creating the Sketch

Create a *Sketch*. Click the **Sketch** icon from the Sketch toolbar.

The 2D profile of the hole is a circle. Create a circle. Click the **Circle** icon from the Sketch Tools toolbar. Create the first point. Click the *Origin*. Create the second point. Drag the **mouse pointer** away from the *Origin*, Figure 1.84. Create the circle. Release the **mouse button**.

Figure 1.84

Click the **Dimension** icon. Click the **circumference of the circle**. Drag the **mouse pointer** directly to the right. Click the **text location**. Enter **10** for the new diameter, Figure 1.85.

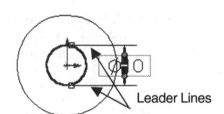

Leader Lines

Figure 1.85

Position the dimension text above the leader lines to improve visibility. Right-click in the **Graphics window**. Click **Select**. Click and drag the dimension **text** upward and to the left, Figure 1.86.

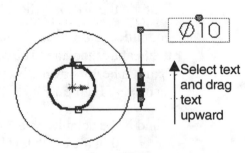

Select text and drag text upward

Figure 1.86

Flip the dimension arrows to the outside of the leader
lines.  Click the **Outside button**
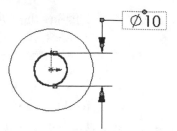 in the PropertyManager.  The
dimension arrows flip to the outside, Figure 1.87.

Figure 1.87

### 1.18.2  Create the Extruded-Cut Feature

•Extrude the *Sketch*.  Click the **Extruded-Cut**
icon on the Features toolbar.  The *Extruded-Cut*
feature dialog box is displayed, Figure 1.88.

Blind is the default Type option.  Enter **10** for the Depth.  Click **OK**, Figure 1.89.

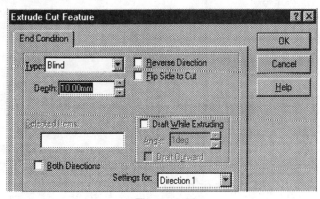

Figure 1.88

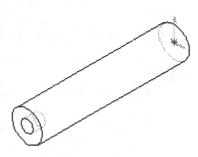

Figure 1.89

The feature is named *Extrude-Cut1*.  Rename ***Extrude-Cut1*** to ***Front-Hole***.

**Save** the ROD.

### 1.19  Chamfer Feature

The *Chamfer* feature removes material along an edge.
The *Chamfer* feature assists the ROD by creating beveled
edges for ease of movement in the GUIDE.

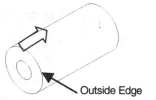
Outside Edge

Figure 1.90

Create a *Chamfer* feature.  Click on the **front outer
circular edge**, Figure 1.90.  An arrow indicates the
direction in which the distance of
the chamfer is measured.

Click the **Chamfer** icon.
The *Chamfer* feature dialog box
is displayed, Figure 1.91.
Enter **2**.  Enter **45** for the Angle.

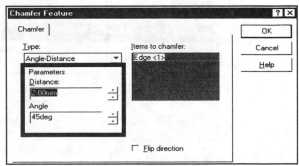

Figure 1.91

Display the *Chamfer* feature, Figure 1.92. Click **OK**.

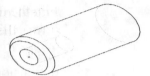

Figure 1.92

## 1.20 View Orientation and Named Views

The View Orientation defines the preset position of the ROD in the graphics window. When creating and editing features, it is helpful to display one particular view or multiple views, Figure 1.93.

The View Orientation options provide the ability to:

- Create a named view.

- Select standard views: *Normal To, Front, Back, Left, Right, Top, Bottom* and *Isometric*.

- Select the *Trimetric* and *Diametric* views.

- Redefine the standard views or return them to the system default setting.

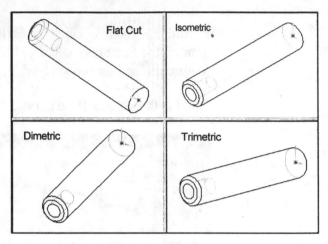

Figure 1.93

Position the ROD. Click the **Rotate** 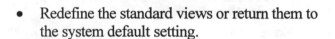 icon. View the back cylindrical surface. Rotate the **ROD**, Figure 1.94.

Create a new View Orientation. Click the **View Orientation** icon from the View toolbar.

Click the **Push Pin** icon from the View toolbar to maintain the displayed menu, Figure 1.95.

Click the **New View** icon in the Orientation menu. Enter *Flat-Cut* in the View Name text box. Click **OK**, Figure 1.96.

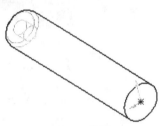

Figure 1.94

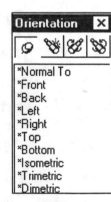

Figure 1.95

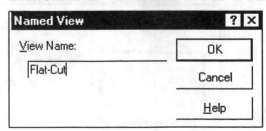

Figure 1.96

Note: Double-click a view name to display the part in a different orientation.  This allows the selection of hidden feature views.

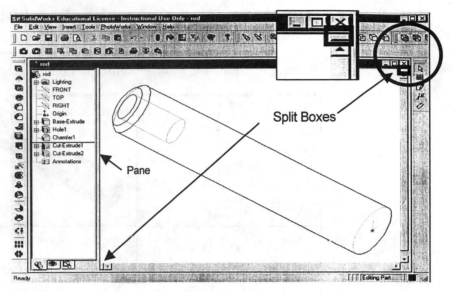

Figure 1.97

Create the vertical panes.  Click the **Split box** at the lower corner of the window, Figure 1.97.  The mouse pointer displays the ←||→ symbol when positioned on the split bar.  Drag the **mouse pointer** to the right.

Create the horizontal panes.  Click the **Split box** in the upper right corner of the window, Figure 1.98.

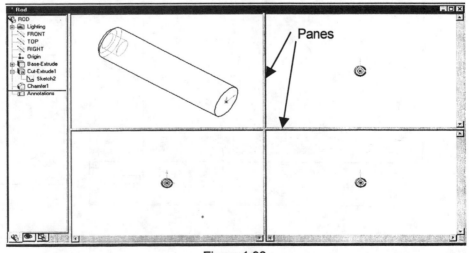

Figure 1.98

Set four views for third angle Orthographic projection. Change the View Orientation. Click inside the **upper right pane**. Click the **Isometric** 🔲 icon. Click inside the **upper left pane**. Click the **Top** 🔲 icon. Click inside the **lower left pane**. Click the **Front** 🔲 icon. Click inside the **lower left pane**. Click **Right** 🔲 icon, Figure 1.99.

Create a single view. Position the **mouse pointer** over the pane bars.

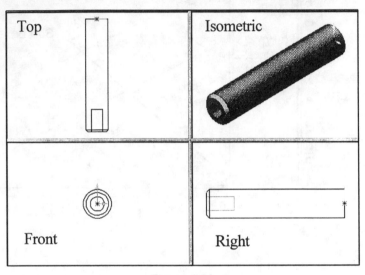

Figure 1.99

Drag the **split bars** to their *Origin*, Figure 1.100. Display the *Flat-Cut* view. Click *Flat-Cut* from the View toolbar, Figure 1.101.

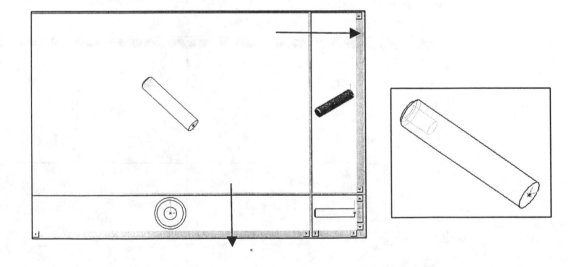

Figure 1.100                                        Figure 1.101

## 1.21 Extruded-Cut Feature

The *Extruded-Cut* feature removes material from the backside of the ROD.  The *Sketch* uses the Convert Entities tool in the Sketch Tools menu to extract existing geometry.

Select the *Sketch* plane.
Click the **back circular face** of the ROD, Figure 1.102.

Click the **Normal To** icon.  The back face of the ROD is displayed, Figure 1.103.

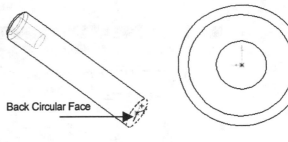

Back Circular Face

**Figure 1.102**

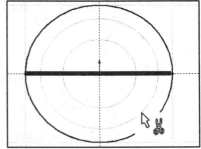

**Figure 1.103**

### 1.21.1 Create the Sketch

Create a *Sketch*.  Click the **Sketch** icon from the Sketch toolbar.  Click the **Convert Entities** icon.

Note: The system extracts the outside edge of the back face and positions it on the *Sketch* plane.

Click the **Line** icon from the Sketch Tools menu.  Sketch a **horizontal line** through the center of the circle, Figure 1.104.  The end points of the line are coincident with the circumference of the circle.

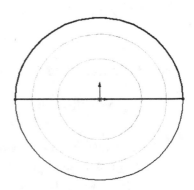

**Figure 1.104**

The Trim command deletes the sketched geometry.  Click the **Trim** icon.  Click the **lower edge** of the circle, Figure 1.105.

**Figure 1.105**

### 1.21.2  Create the Extruded-Cut Feature

Extrude the *Sketch*.  Click the **Extruded-Cut** ⬛ icon on the Features toolbar. The *Extruded-Cut* feature dialog box is displayed, Figure 1.106.  Click the ***Flat-Cut*** view.

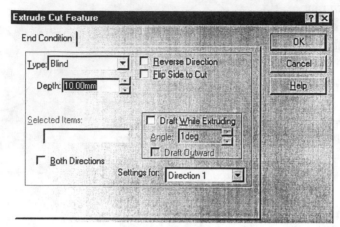

Figure 1.107

Figure 1.106

Enter **10** for the Depth.  Display the *Extruded-Cut*, Figure 1.107.  Click **OK**.

Rename ***Cut-Extrude2*** to ***Flat-Cut***.

Orient the part in the *Isometric* view. Click the **Isometric** 🔲 icon from the Standards View toolbar, Figure 1.108.

**Close** the View Orientation dialog box.

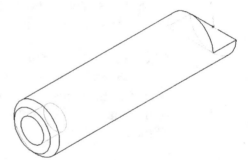

Figure 1.108

### 1.22  Move/Size an Extruded Feature

Modified the *Extruded* feature by using the *Move/Size* feature handles.

Resize the *Flat-Cut* feature.  Click the **Resize** 🔼 icon from the Features toolbar.  Click the ***Flat-Cut*** feature from the FeatureManager.  The *Move/Size* handles are displayed on the *Flat-Cut* feature, Figure 1.109.

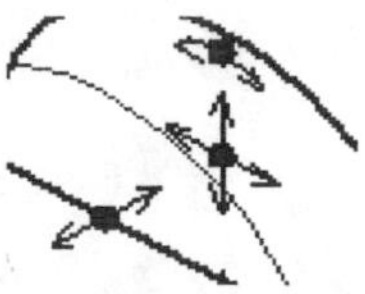

Figure 1.109

Click the **Move** handle. Drag the **Move** handle toward the *Front-Hole*, Figure 1.110. Release the **mouse button** when 30 is displayed on the pointer, Figure 1.111.

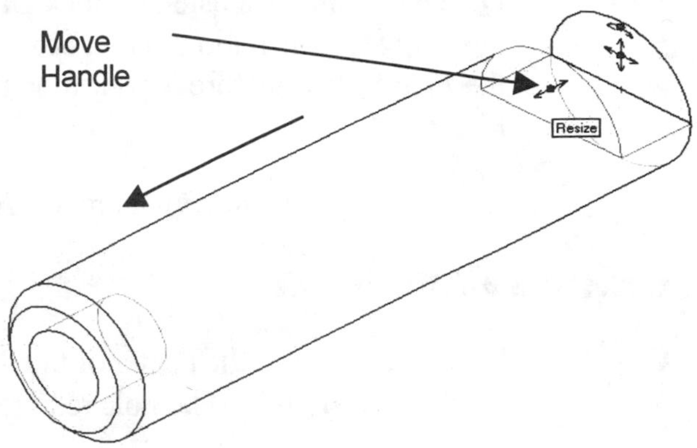

Turn off the **Resize** command. Click the **Resize** ⇕ icon.

Rebuild the part. Click the **Rebuild** icon, Figure 1.112.

**Save** the ROD.

## 1.23 Copy/Paste Function

The Copy/Paste function provides the ability to copy and paste information.

The ROD requires an additional hole on the flat face.

Copy the *Front-Hole* feature to the flat face.

Click the *Front-Hole* feature. Click the **Copy** icon. Click the **center of the flat face**.

Click the **Paste** icon, Figure 1.113.

Note: The Copy Confirmation dialog box appears automatically. The box states there are external constraints in the feature being copied, Figure 1.114. External constraints are the dimensions used to place the *Front-Hole*.

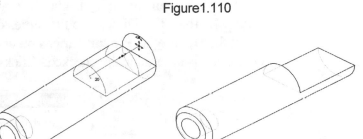

Figure1.110

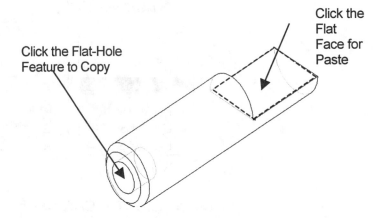

Figure 1.111                    Figure 1.112

Click the Flat-Hole Feature to Copy

Click the Flat Face for Paste

Figure 1.113

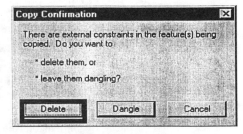

Figure 1.114

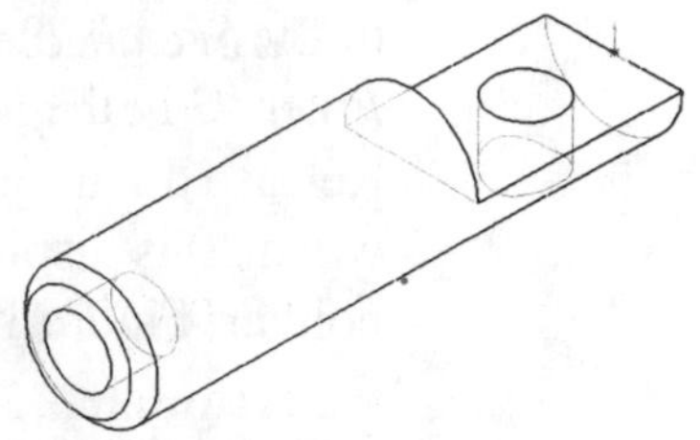

Delete the old dimensions. Click **Delete** from
the Copy Confirmation dialog box. A copy of
the *Front-Hole* feature is placed on the flat face,
Figure 1.115.

Rename *Cut-Extrude3* feature to *Flat-Hole*.

### 1.23.1 Create Dimensions

Figure 1.115

Locate the hole in the center of the flat face. Right-click the *Flat-Hole* feature.
Click **Edit Sketch**. The hole requires two dimensions.

Modify the dimension style. Right-click on the dimension **text**. Click Properties,
Figure 1.116a. The Dimension Properties dialog box is displayed, Figure 1.116b.
Click the **Display as Linear dimension** check box. The Diameter dimension
check box is automatically checked. Click **OK**.

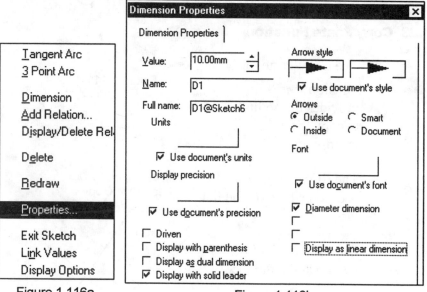

Figure 1.116a          Figure 1.116b

Create the first dimension. Click the **Dimension** [icon] icon.

Click the **back edge** of the ROD. Click the **center point of the circle**,
Figure 1.117a.

Figure 1.117a          Figure 1.117b

Place the **dimension** to the right of the *Sketch*, Figure 1.117b. Enter **15** in the Modify dialog box.

Add geometric relations. Click the **Geometric Relations** icon from the Sketch Relations toolbar. Click the **center point** of the circle. Click the *Origin*, Figure 1.118.

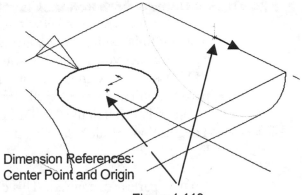

Dimension References: Center Point and Origin

Figure 1.118

Align the *Origin* and center point of the circle. Click **Vertical** from the Add Geometric Relations dialog box, Figure 1.119a. Click **Apply**. Click **Close**.

Display the Top view. Click the **Top** icon. The *Flat-Hole* feature is vertically aligned to the *Origin*, Figure 1.119b.

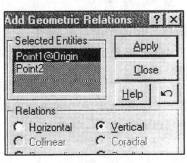

Figure 1.119a                    Figure 1.119b

The *Sketch* is fully defined. Close the *Sketch*. Click the **Sketch** icon.

The *Flat-Hole* feature requires a drilled hole. Edit the feature definition. Right-click on the *Flat-Hole* feature. Click **Edit Definition** from the Pop-Up menu.

Click the **drop down arrow** from the Type list box. Click the **Through All** option, Figure 1.120. Display the updated *Flat-Hole* feature. Click **OK**, Figure 1.121.

**Save** the ROD.

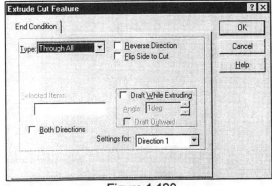

Figure 1.120

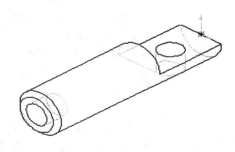

Figure 1.121

## 1.24  Design Change with Rollback and Edit Definition

You are finished for the day.  The phone rings.
The customer voices concern with the GUIDE-
ROD assembly.  Can you create an additional
part that will prevent the ROD from moving
completely through the GUIDE?  You provide
a suggestion.  Build a stop into the ROD,
Figure 1.122.

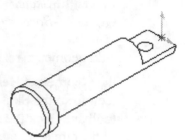

No new parts are required.  This design will
control cost.  The customer agrees but wants to
view a copy of the GUIDE and ROD design by tomorrow!

Figure 1.122

You are required to implement the design change and to incorporate it into the
existing part.  Changes are not required for the GUIDE.  You begin the design
change for the ROD.

The Rollback and Edit Definition functions are used to
implement the design change.

The Rollback function allows a feature to be redefined in
any state or order.  The Edit Definition function allows
feature parameters to be redefined.

Implement the design change.  First, add the new *Extruded-
Boss* feature to the front face of the ROD.  Second, delete
the first *Front-Hole* feature.

Note: In this procedure you are exposed to rebuild errors.
Information is provided to correct these errors.

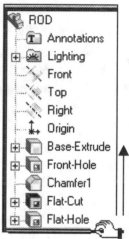

Rollback Bar
Figure 1.123

Create the *Extruded-Boss* feature before the *Front-Hole*
feature.  Place the **mouse pointer** over the yellow Rollback
bar at the bottom of the FeatureManager design tree,
Figure 1.123.  The mouse pointer displays a symbol of a hand.

Drag the **Rollback** bar upward to
below the *Base-Extrude* feature.
Only the *Base-Extrude* feature is
displayed, Figure 1.124.  The
Rollback bar is blue.

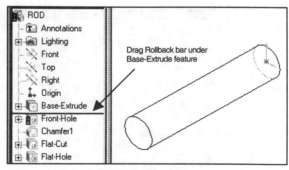

Figure 1.124

Create a new *Extruded-Boss* feature.

Select the *Sketch* plane. Click the **front face** of the
*Extruded-Base* feature, Figure 1.125.

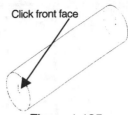

Figure 1.125

Create a *Sketch*. Click the **Sketch** icon from the
Sketch toolbar. Create a circle. Click the **Circle**
icon. Create the first point.

Click the ***Origin*** of the *Extruded-Base*
feature. The mouse pointer displays a pencil
with a point .

Create the second point. Drag the **mouse
pointer** away from the *Origin*. Release the
**mouse button**.

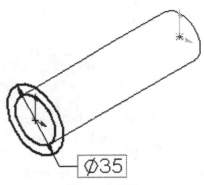

Add a dimension. Click the **Dimension**
icon. Click the **circumference**.
Enter **35**, Figure 1.126.

Figure 1.126

Note: The *Origin* is located on the back face. The *Origin* is projected to the front
face in the *Isometric* view. No dimension is required for the inside circle. The
inside circle dimension is determined from the *Extruded-Base* edge.

### 1.24.1  Extrude the Sketch

Extrude the *Sketch*. Click the
**Extruded Boss/Base** icon on
the Features toolbar. The Extrude
Feature dialog box is displayed,
Figure 1.127.

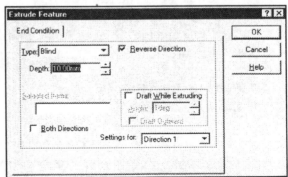

Create the feature in the direction
of the flat face. Click the **Reverse
Direction** Check box. Blind is
the default Type option. Enter **10** in the
Depth list box.

Figure 1.127

Display the *Extruded-Boss* feature,
Figure 1.128. Click **OK**.

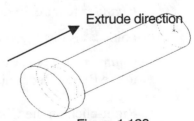

Figure 1.128

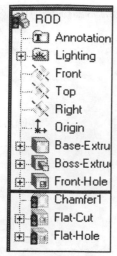

Figure 1.129

The feature is named *Boss-Extrude1*.

Drag the **Rollback** bar downward below the *Front-Hole* feature and above the *Chamfer1* feature, Figure 1.129.

The *Front-Hole* feature is displayed, Figure 1.130.

The *Front-Hole* feature is no longer required. Delete the *Front-Hole* feature. Click the ***Front-Hole*** feature. Right-click and click **Delete** from the Pop-up menu.

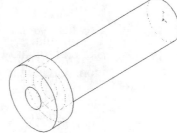

Figure 1.130

The Confirm Delete dialog box appears, Figure 1.131. *Sketch2* is the circular profile used to create the *Front-Hole*. Delete the dependent sketched feature. Click the **Also delete absorbed features** text box. Click **Yes**. The feature is removed from the ROD and the FeatureManager.

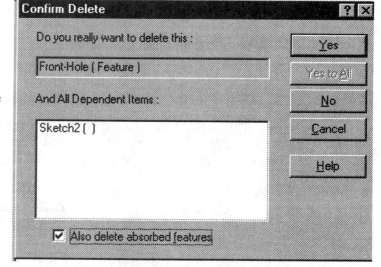

Figure 1.131

### 1.24.2  Recover from Rebuild Errors

Rebuild errors can occur when using the Rollback function. A common error occurs when an edge or face is missing.

Drag the **Rollback** bar downward below the *Chamfer* feature, Figure 1.132. A red explanation point ⚠ is displayed next to the name of the *Chamfer* feature. A red arrow is displayed 🔴 next to the ROD. Red indicates a model Rebuild error.

Right-click on the **Chamfer** name. Click **What's Wrong**?

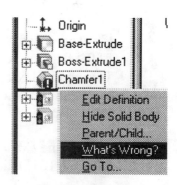

Figure 1.132

The ROD Rebuild Error dialog box is displayed, Figure 1.133. Click **Close**.

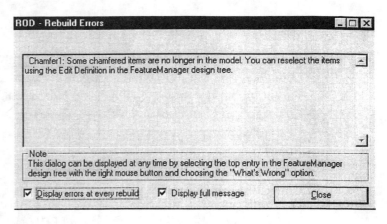

Figure 1.133

The original edge from the *Extruded-Base* feature was deleted when the *Boss-Extrude1* feature was created.

Create the *Chamfer* feature on the *Boss-Extrude1* edge. Right-click the **Chamfer1** in the FeatureManager. Click **Edit Definition** from the Pop Up menu.

Figure 1.134

The Caution dialog box displays the message, "Chamfer1 is missing 1 edge", Figure 1.134. Click **OK**.

The Chamfer Feature dialog box appears, Figure 1.135.

Click the **front edge** of the *Boss-Extrude1* feature, Figure 1.136. The item

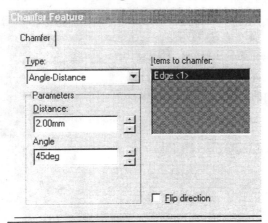

Figure 1.135

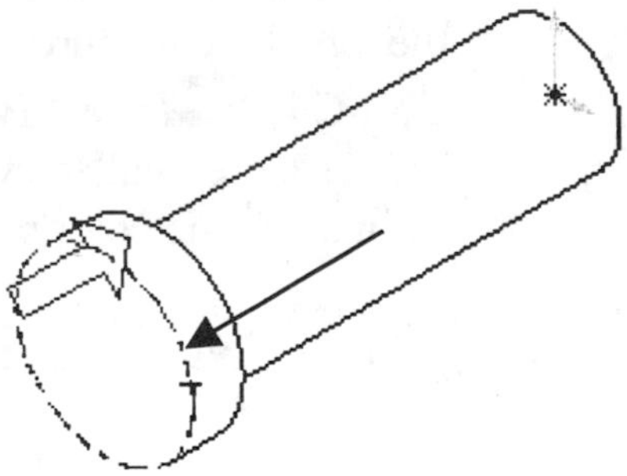

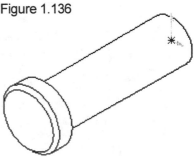

Figure 1.136

appears inside the Items to Chamfer list box.

Create the *Chamfer* feature. Click **OK** from the Chamfer Feature dialog box, Figure 1.137.

Figure 1.137

### 1.24.3 Complete the Rollback

Check to insure that the part is completely rebuilt. Drag the **Rollback** bar downward to the bottom of the FeatureManager, Figure 1.138.

All feature icons in the FeatureManager design tree are in yellow.

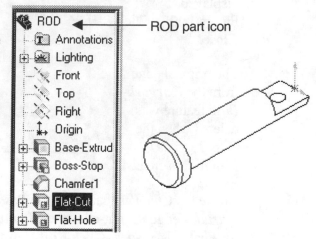

Figure 1.138

Rename *Boss-Extrude1* to *Boss-Stop*.

Modified feature names or dimensions are displayed in blue. Rebuild all parts. Click the **Rebuild** icon from the Main toolbar. Save the **ROD**.

### 1.24.4 Edit Part Color

All parts are shaded gray by system default. Modify the system default color. Click the **ROD Part** icon on the top of the FeatureManager design tree. Click the **Edit Color** icon. Basic colors are displayed as small squares, called swatches. Click a **color swatch** from the Basic color palette, Figure 1.139.

Set part color. Click **OK** from the Edit Color dialog box.

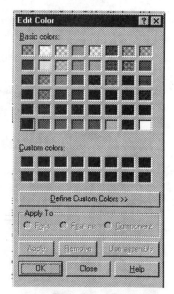

Figure 1.139

Display the color. Click the **Shade**  icon from the View toolbar, Figure 1.140.

**Save** the ROD.

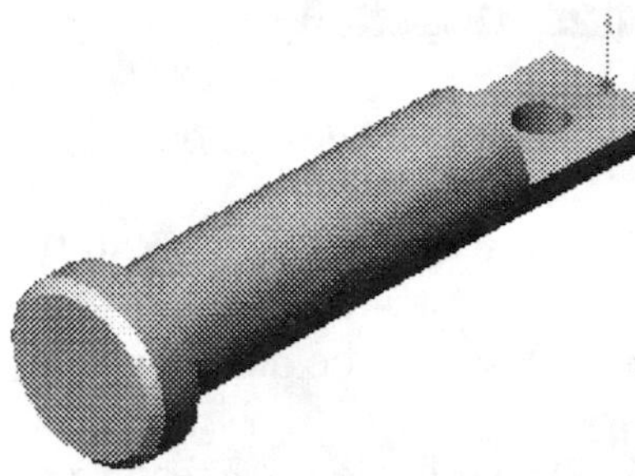

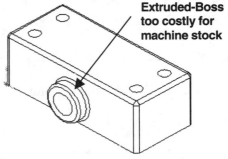

Figure 1.140

Editing features is as important as creating features. Design changes are an integral part of the engineering process. In Project 2, you will incorporate the ROD and GUIDE into an assembly. After the initial assembly, you will modify the ROD to complete the customer requirements.

Project 1 is completed. The parts are similar to your original concept *Sketch*. You performed design changes to support the requirements of the customer and to address manufacturing and serviceability issues. The conceptual design is filled with changes.

Can you incorporate other changes to these parts? Are there any sharp edges?

Create numerous features in SolidWorks. Example: The *Extruded-Boss*, Figure 1.141, is created by converting the inside *GuideHole* edge.

**Extruded-Boss too costly for machine stock**

Remember, ask questions! How would this new feature change the part? The *Extruded-Boss* would be very costly to manufacture from machined bar stock. As a designer, review the manufacturing options, before you start the design.

Figure 1.141

For machined parts, begin with a *Extruded-Base* feature that represents standard machine stock. Examples of standard machine stock are provided in the following exercises.

Are you ready to start Project 2? Stop! Let's examine and create additional parts. Perform design changes. Take chances, make mistakes and have fun with the various features and commands.

ROD

## 1.25 Questions

1. What key design questions should you investigate before starting a design?

2. How do you start a SolidWorks part?

3. Describe the default reference planes.

4. What is the *Base* feature and why should the *Base* feature be kept simple?

5. What is the difference between an *Extruded-Base* feature and an *Extruded-Cut* feature?

6. What is a *Round* feature?

7. What command keys are used to *Move* and *Copy* sketched geometry?

8. Describe the *Edit* feature.

9. What is a *Chamfer* feature?

10. Describe the Rollback function.

11. What type of Geometric Relations can you add to a *Sketch*?

12. Describe the *Move/Size* feature.

## 1.26 Exercises

### I. Create the following Extruded Parts:

Exercise 1.1        ROUND-BAR, Hot Rolled Round Bar Steel Stock
Exercise 1.2        FLAT-PLATE, Cold Finished Flat Steel Stock
Exercise 1.3        L-BRACKET, Cast Iron
Exercise 1.4        U-SECTION Aluminum

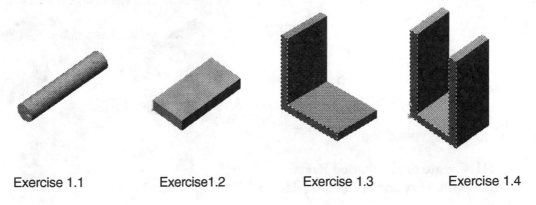

Exercise 1.1          Exercise1.2          Exercise 1.3          Exercise 1.4

Exercise 1.5        T-SECTION, Cast Iron
Exercise 1.6        V-SECTION, Cast Iron
Exercise 1.7        BLOCK, Cast Iron

Exercise 1.5          Exercise 1.6          Exercise 1.7

Parts courtesy of
Reid Tool Supply Co.
Muskegon, MI USA

Parts manufactured from machine stock can save design time and money.  Machine
stock is purchased from material supply companies,
Example: Reid Tool Supply Company (www.reidtool.com).

## II. Create the Extruded Parts.

State the Sketch Plane for each part based upon the orientation displayed.

Exercise 1.8     LOCATING PIN Sketch Plane _____

Exercise 1.9     BUSHING Sketch Plane_____

Exercise 1.10    KEYWAY BUSHING Sketch Plane _____

Exercise 1.8 LOCATING PIN       Exercise 1.9 BUSHING     Exercise 1.10 KEYWAY BUSHING

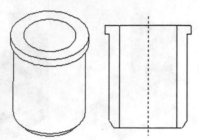

## III. Create the Extruded Parts.

Use symmetry and geometric relations.

Exercise 1.11    JOINING –STRIP

Exercise 1.12    TEE-JOINING-PLATE

Exercise 1.13    90DEGREE-JOINING-PLATE

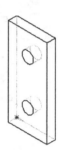

Exercise 1.11

The company, 80/20 Inc., manufactures modular aluminum structural extrusions to reduce fabrication time and cost. 80/20 manufacturers both Metric and English products.

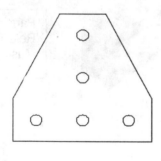
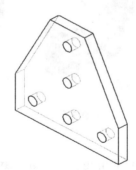

Exercise 1.12

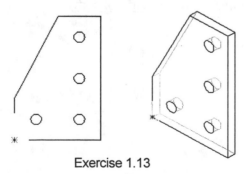

Exercise 1.13

Parts courtesy of 80/20, Inc.
Columbia City, IN USA
(www.8020.net)
and its distributor, Air Inc.
Franklin, MA USA
(www.airinc1.com)

## IV. Create the Extruded Parts.

Parts are created from machine stock. No dimensions are given.

Exercise 1.14 Create one STANDOFF.  Create one PLATE.
             Four STANDOFFs will fit between two PLATEs (Exercise 2.2).

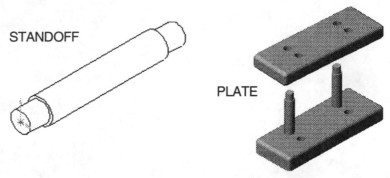

STANDOFF

PLATE

Exercise 1.14

Exercise 1.15 Create a PIN.  Create a U-BRACKET.

Exercise 1.16 Create a T-SECTION and MOUNTING PLATE.

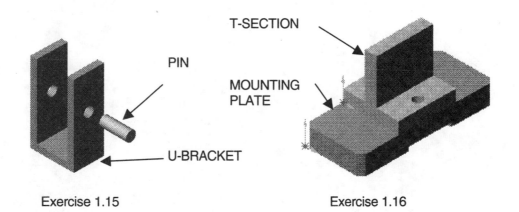

PIN

T-SECTION

MOUNTING
PLATE

U-BRACKET

Exercise 1.15                                    Exercise 1.16

## V. Create part document template.

Create a Metric part document template using an ISO dimension standard.

Create an English part document template using an ISO dimension standard.

Set Drawing/Units to appropriate values for each template.

Notes

# Project 2

## Fundamentals for Assembly Modeling

Below are the desired outcomes and usage competencies based upon the completion of Project 2.

| Project Desired Outcomes: | Usage Competencies: |
|---|---|
| • Knowledge and experience in assembly creation. | • Ability to create Sketched Geometry, features and Dimensional schemes. |
| • A comprehensive understanding of incorporating design changes into an assembly. | • Ability to Modify, Edit, and Redefine assembly features. |
| • A GUIDE-ROD assembly. | • Ability to comprehend the function of assembly creation. |

# 2   Project 2 – Fundamentals of Assembly Modeling

## 2.1   Project Objective

Create a GUIDE-ROD assembly.

## 2.2   Project Situation

In Project 1, you created two parts:

1. ROD

2. GUIDE

Now, you are required to assemble the parts and create an assembly.  First, let's review the assembly design constraints:

- The ROD requires the ability to travel through the GUIDE.

- The flat end of the ROD must remain parallel to the top surface of the GUIDE. Note: Install the GUIDE so the top surface is parallel to the work area.

- The ROD requires a minimum linear travel of 9 centimeters through the GUIDE.


Rough Sketch of Design Situation

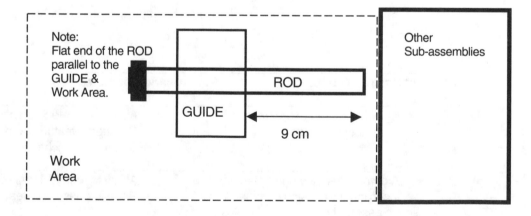

Figure 2-1

Create a rough sketch of the conceptual assembly, Figure 2.1.

An assembly or sub-assembly combines two or more parts. In an assembly or sub-assembly parts are referred to as components.

Design constraints directly influence the assembly design process. Other considerations indirectly impact the assembly design, namely: cost, manufacturability and serviceability.

## 2.3    Project Overview

Translate the rough conceptual sketch into a SolidWorks assembly. Determine the first component of the assembly. The first component is the GUIDE.

The action of assembling components in SolidWorks is defined as Mates, Figure 2.2. Mates are relationships between components that simulate the construction of the assembly in a manufacturing environment.

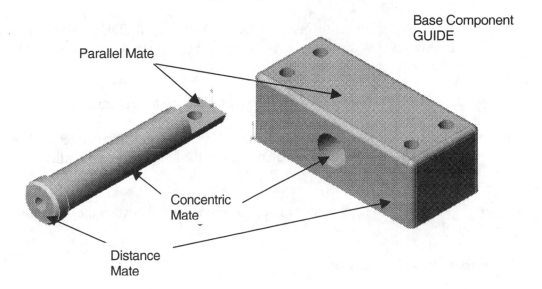

Figure 2-2

## 2.4    Tolerance and Fit

The ROD travels through the GUIDE in the GUIDE-ROD assembly. The cylindrical shaft diameter of the ROD is 25 mm. The hole diameter in the GUIDE is 25 mm. A 25 mm ROD cannot be inserted into a 25 mm GUIDE hole without great difficulty! Note: The 25 mm dimension is the nominal dimension. The nominal dimension is approximately the size of a feature which corresponds to a common fraction or whole number.

Tolerance is the difference between the maximum and minimum variation of a nominal dimension and the actual manufactured dimension.

Example: A ROD has an nominal dimension of 100 mm with a tolerance of ± 2 mm, (100 mm ± 2 mm). This translates to a part with a possible manufactured dimension range between 98 mm to 102 mm. The total ROD tolerance is 4 mm.

Note: Design rule of thumb: Design with the maximum permissible tolerance. Tolerance flexibility saves in manufacturing time and cost.

The assembled relationship between the ROD and the GUIDE is called the fit. The fit is defined as the tightness or looseness between two components. There are three major types of fits:

- Clearance fit - The shaft diameter is less than the hole diameter.

- Interference fit – The shaft diameter is larger than the hole diameter. The difference between the shaft diameter and the hole diameter is called interference.

- Transition fit – Clearance or interference can exist between the shaft and the hole.

## 2.5    Assembly Modeling Approach

In SolidWorks, components and their assembly are directly related through a common database. Changes in the components directly affect the assembly and vise a versa. SolidWorks provides two assembly-modeling techniques:

- Top down

- Bottom up

### 2.5.1  Top Down Design Approach

In the Top down approach, major design requirements are translated into assemblies, sub-assemblies and components. Note: To start, you do not need all of the required component design details. Individual relationships are required.

Example: A computer is divided into key sub-assemblies such as a: motherboard, disk drive and power supply, Figure 2.3.

Figure 2.3

Relationships between these sub-assemblies must be maintained.

### 2.5.2  Bottom Up Design Approach

In the Bottom up approach, components are assembled using part dependencies and parent-child relationships. Note: In this approach, you possess all of the required design information for the individual components.

The child is automatically modified during a parent modification. Avoid unwanted references and dependencies when establishing parent-child relationships.

The Bottom up approach is used in this project for the GUIDE-ROD assembly.

## 2.6  Linear Motion and Rotational Motion

In dynamics, motion of an object is described in linear and rotational terms. Components possess linear motion along the x, y and z-axes and rotational motion around the x, y, and z-axes. In an assembly, each component has 6 degrees of freedom: 3 translational (linear) and 3 rotational. Mates remove degrees of freedom. All components are rigid bodies. The components do not flex or deform.

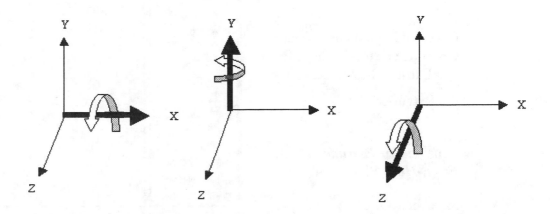

### 2.7    Verify and Modify the ROD and GUIDE

Before you create an assembly, verify the creation of the parts. You can add new features or modify existing features at any time.

Open the ROD. Click the **Open** 📂 icon. Enter the part name. Enter **ROD**.

Review each feature. Display the *Base-Extrude* feature. Drag the **Rollback** bar upward to the top of the FeatureManager. Drag the **Rollback** bar downward, one feature at a time, until you reach the bottom of the FeatureManager.

Add or modify features. Create a new hole on the front face of the ROD, Figure 2.4a. Do you remember the required steps?

- Select the *Sketch* plane

- Create the *Sketch*

- Dimension the *Sketch*

- Add Relations if required

- Extrude the *Sketch*

Add an *Extruded-Cut* feature to the ROD.

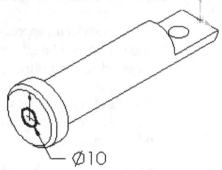

Ø10

Figure 2.4a

Select the *Sketch* plane. Click the **front face**.

Create a *Sketch*. Click the **Sketch** ✏️ icon from the Sketch toolbar.

The 2D profile of the hole is a circle. Create a circle. Click the **Circle** ⊙ icon from the Sketch Tools toolbar. Create the first point. Click the *Origin* ↳. Create the second point. Drag the **mouse pointer** away from the *Origin*. Create the circle. Release the **mouse button**.

Click the **Dimension** 🖉 icon. Click the **circumference**. Enter **10**. Extrude the *Sketch*. Click the **Extruded Cut** 🖻 icon. Enter **10** for Depth.

Rename *Cut-Extrude* to *Front-Hole*, Figure 2.4b.

**Save** the ROD.

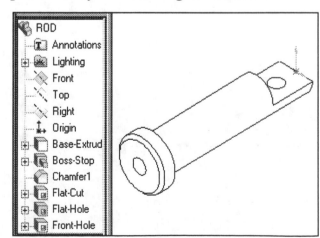

Figure 2.4b

Modify the overall length of the ROD. How do you determine the minimum ROD length required? Estimate a 150 mm's, based upon the customer requirement and the depth of the GUIDE.

90 mm (travel distance) + 60 mm (GUIDE depth) = 150mm(ROD length)

Double-click the **shaft** of the ROD. The dimensions of the *Base-Extrude* feature are displayed. Double-click on the **120** text. Enter **150** in the Modify dialog box. Click **Rebuild**, Figure 2.4c.

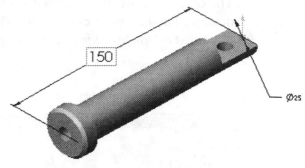

Verify and review each feature of the GUIDE. Open the GUIDE. Click the **Open** 📂 icon. Enter **GUIDE**.

Figure 2.4c

Review each feature. Drag the **Rollback** bar upward to the top of the FeatureManager. Drag the **Rollback** bar downward, one feature at a time until you reach the bottom of the FeatureManager.

Color aids in assembly visualization. Change the part color. Click **Edit Color** 🔲 icon. Select a **color** (not gray). Click **OK**.

**Save** the GUIDE.

## 2.8    Create an Assembly

Create the GUIDE-ROD assembly. Click the **New** 🗋 icon from the Standard toolbar. The New SolidWorks Documents dialog box is displayed.

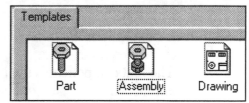

Figure 2.5a

Click **Assembly** from the Templates folder, Figure 2.5a. Click **OK**. The default file name is ASSEM1. Click **File** from the Main menu. Click the **Save** 💾 icon. Enter the assembly name.

Enter **GUIDE-ROD**, Figure 2.5b.

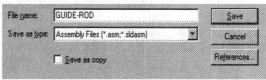

Figure 2.5b

Set assembly units. Click **Tools** from the Main menu. Click **Options**. Click the **Document Properties** tab. Click **Units**. Click **Millimeters** from the Linear units drop down list. Click **OK**.

Verify that the ROD and GUIDE are open. Click **Window** from the Main menu. The following components are displayed: ROD, GUIDE and GUIDE-ROD. Display the assembly and components. Click **Tile Horizontally** from the Window menu, Figure 2.5c.

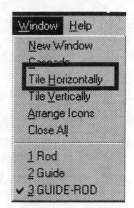

The *Origin* feature of the GUIDE-ROD assembly is displayed in the Graphics window. The FeatureManager is displayed on the left side of the three graphics windows, Figure 2.6

Figure 2.5c

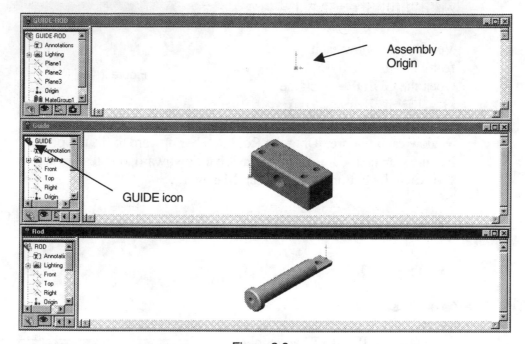

Figure 2.6

### 2.8.1   First Component

The first component is the foundation of the assembly. The GUIDE is the first component.

Components are added to the assemblies utilizing the following techniques:

- Click Insert from the Main menu. Click Component. Click From File.

- Drag components from Windows Explorer.

- Drag components from the Feature Palette window.

- Drag components from the Open part files.

Position the GUIDE inside the assembly Graphics window.

Click the **GUIDE** 🖘 GUIDE    icon from the top of the FeatureManager. Drag the **GUIDE** to the *Origin* ↳ of the GUIDE-ROD assembly window.

The mouse pointer displays ⬚ when position on the *Origin*. Display the GUIDE component in the assembly Graphics window. Release the **mouse button**, Figure 2.7a.

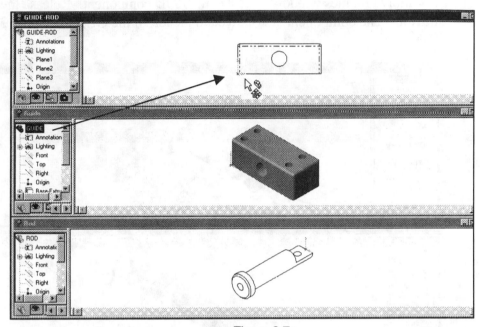

Figure 2.7a

The GUIDE name is added to the GUIDE-ROD assembly FeatureManager with the symbol (f). The symbol (f) represents a fixed component. A fixed component cannot move and is locked to the assembly *Origin*.

Close the GUIDE. Click the **Close** 🗙 icon. Resize the windows. Drag the top horizontal border of the ROD window upward. Drag the bottom horizontal border of the GUIDE-ROD window downward, Figure 2.7b.

Display the GUIDE inside the assembly window. Click inside the **GUIDE-ROD** window. Display the *Isometric* view of the GUIDE. Click the **Isometric** 🔲 icon.

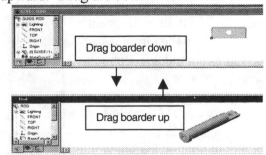

Figure 2.7b

### 2.8.2   Second Component

The second component is the ROD.  Position the ROD component inside the assembly window.

Click the **Isometric** icon in the GUIDE-ROD window.  Click the **ROD** icon from the top of the FeatureManager.  Drag the **ROD** component to the assembly window.

The mouse pointer displays when positioned inside the GUIDE-ROD assembly graphics window.  Display the ROD component.  Release the **mouse button**.  The component is displayed to the left of the GUIDE, Figure 2.8.

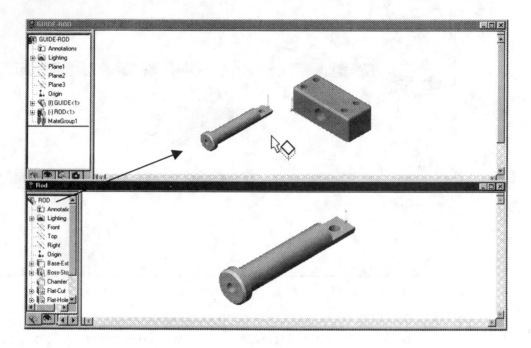

Figure 2.8

Enlarge the assembly window.  Click the **Maximize** icon in the upper right hand corner of the GUIDE-ROD window, Figure 2.9.

Fit all components
in the Graphics
window.  Press the
**f** key.

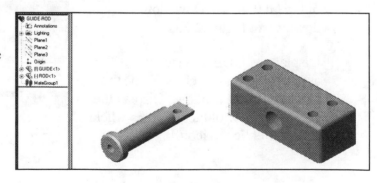

Figure 2.9

Save the GUIDE-ROD assembly.  Click the **Save** icon.

### 2.8.3  FeatureManager Syntax

Double click the **GUIDE** component inside the
FeatureManager.  The GUIDE component displays all
features, Figure 2.10.

Let's review the ROD
syntax components in the
FeatureManager.

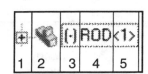

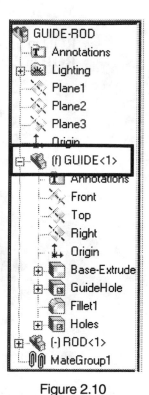

1.  A Plus sign ⊞ icon
    indicates that additional feature information is
    available.  A minus sign indicates that the feature list
    is fully expanded.

2.  A component icon indicates that the ROD is a

    part.  The assembly icon indicates that the
    GUIDE-ROD is an assembly.  Sub-assemblies
    display the same icon as an assembly.

3.  Column 3 identifies the Component State:

    Figure 2.10

    a.  A minus sign (–) indicates that the
        component is under defined and requires additional information.

    b.  A plus sign (+) indicates that the component is over defined.

    c.  A fixed symbol (f) indicates that the component does not move.

    d.  A question mark (?) indicates that additional information is required.

4.  ROD - Name of component.

5.  The symbol <#> indicates the number of copies in the assembly.  The symbol
    <1> indicates the original component, "ROD" in the assembly.

### 2.8.4   Move Component

Components located in an assembly are free to move.  Move the ROD.  Click the **Move Component** icon from the Assembly toolbar.  Click the **shaft** of the ROD in the Graphic Window.  The mouse pointer displays . Position the **ROD** in front of the GUIDE hole, Figure 2.11.

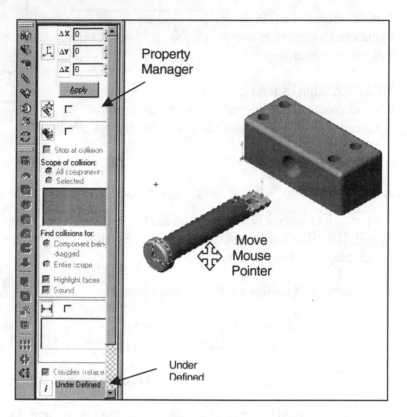

Figure 2.11

The PropertyManager is displayed on the left side of the Graphics window.

Under Defined is displayed at the bottom of the PropertyManager menu.  An under defined component requires additional information.

### 2.8.5   Mate Component

Mates are relationships that align and fit components in an assembly.  You will create three Mates in this section.

Components are inserted into an assembly with various intuitive options: Coincident, Parallel, Tangent, Concentric, Distance, Angle and Perpendicular.

Establishing the correct component relationship in an assembly requires forethought on component interaction.

Recall the initial assembly design constraint.  The ROD requires the ability to travel through the GUIDE.

Create the first Mate.  Click the **Mate** icon from the Assembly toolbar.  The Assembly Mating dialog box is displayed, Figure 2.12.  The cylindrical face of the ROD is currently selected.  Click the **cylindrical face** of the GUIDE hole, Figure 2.13.

The faces are added to the Items Selected list.  Click **Concentric** from the Mate Types box.  Click **Closest** from the Alignment Condition text box.  Click **Preview**.  Click **Apply**.

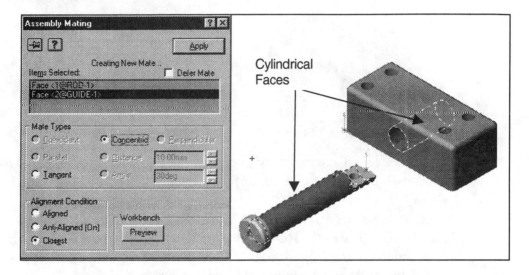

Figure 2.12                                                    Figure 2.13

Note: When selecting faces, position the mouse pointer in the middle of the face.  Do not position the pointer near the edge of the face.  If the wrong face or edge is selected, click the face or edge again to remove it from the Items Selected text box.

Right-click in the Graphics window.  Click Clear Selections to remove all geometry from the Items Selected text box.

Select the hidden geometry with the SelectOther option.  Right-click the

Select icon from the Pop-up menu.  Right-click N until the correct geometry is selected.  Accept the geometry.  Click Y.

The ROD is positioned inside the GUIDE. The ROD has the ability to rotate while remaining concentric to the GUIDE hole.

Click the **Move Component** 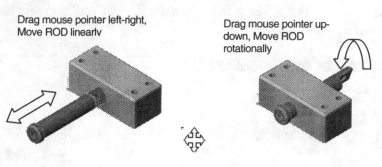 icon from the Assembly toolbar. Drag the **Move Component** icon in a horizontal direction. The ROD travels linearly in the GUIDE, Figure 2.14. Drag the **Move Component** icon in a vertical direction. The ROD rotates in the GUIDE, Figure 2.15.

Drag mouse pointer left-right, Move ROD linearly

Drag mouse pointer up-down, Move ROD rotationally

Figure 2.14　　　　　　　　　　　　　　Figure 2.15

Recall the second assembly design constraint. The flat end of the ROD must remain parallel to the top surface of the GUIDE.

Create the second Mate. Click the **Mate** icon from the Assembly toolbar. Click the **flat face** of the ROD. Click the **flat top face** of the GUIDE. Both faces are added to the Items Selected list. Click **Parallel** from the Mate Types box. Click **Closest** from the Alignment Condition text box, Figure 2.16.

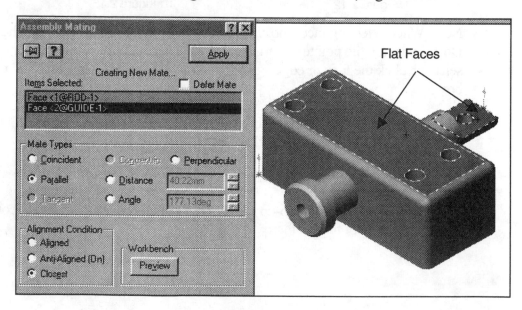

Figure 2.16

Click **Preview**. Click **Apply**.

The ROD continues to travel linearly through the hole. Click the **Move Component**  icon from the Assembly toolbar. Position the **ROD** approximately in the center of the GUIDE. The component remains under defined.

Recall the third assembly design constraint. The ROD requires 9 centimeters of linear travel through the GUIDE.

Position the assembly for the third Mate. Click the **Rotate** icon from the View toolbar. Position the **assembly** to view the backside of the ROD and GUIDE, Figure 2.17.

Figure 2.17

Zoom in on the backside of the ROD and GUIDE. Click the **Zoom to Area** icon.

Selection Filters are used to select difficult individual features such as: faces, edges and points.

Click **View** from the Main menu. Click **Toolbars**. Click **Selection Filter**. Click the **Face** icon from the Selection Filter toolbar, Figure 2.18.

Figure 2.18

Note: When the Face Filter is activated, only component faces can be selected.

Create the third Mate. Click the **Mate** icon button from the Assembly toolbar. Click the **back semi circle face** of the ROD. Click the **back flat face** of the GUIDE, Figure 2.19.

The faces are added to the Items Selected list. Click **Distance** from the Mate Types box. Enter **90** mm. Click **Closest** from the Alignment Condition text box. Click **Apply**.

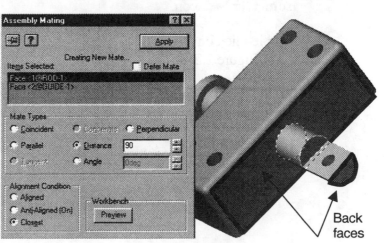

Back faces

Figure 2.19

Remove all Selections Filters. Click the **Clear All Filter** icon.

Display the *Isometric* view. Click the **Isometric** icon.

You have just created three Mates. The ROD component is fully defined in the Assembly. | ⊞ ROD<1> .

The Mates are listed in *MateGroup1*. *MateGroup1* is located in the FeatureManager. Display the Mate types. Double-click on *MateGroup1*, Figure 2.20. Display the full mate names. Drag the vertical FeatureManager border to the right.

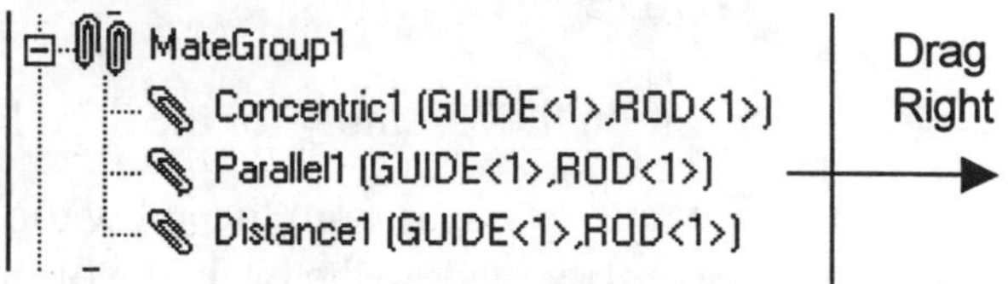

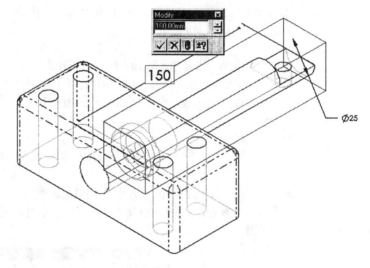

⊟ MateGroup1
  Concentric1 (GUIDE<1>,ROD<1>)
  Parallel1 (GUIDE<1>,ROD<1>)
  Distance1 (GUIDE<1>,ROD<1>)

Drag
Right
→

Figure 2.20

## 2.8.6 Edit Component Dimension

You realize from the assembly that the ROD requires additional length. Modify the feature dimensions in the assembly.

Modify the shaft of the ROD. Click the **Isometric** icon.

Double-click on the **shaft**, Figure 2.21a.

150

Ø25

Figure 2.21a

Modify the 150 dimension. Double-click **150**. Enter **180**.

Update the assembly. Click the **Rebuild** icon, Figure 2.21b.

Figure 2.21b

## 2.9    Add a Library Component

A parts library contains components used in a design creation. Your company issued a design policy. The policy states that you are required to only use parts that are presently in the company's parts library. The policy is designed to lower inventory cost, purchasing cost and design time. What does this mean to you? It means that you cannot use your selected cap head screw.

In this project, the SolidWorks Feature Palette simulates your company's part library. Replace the cap head screw with a hex flange bolt. The hex flange bolt is located in the FeatureLibrary. Modify the hole diameters of the GUIDE to accommodate the hex flange bolt. Note: The original GUIDE holes were created with the Copy command.

### 2.9.1    Feature Palette

The SolidWorks Feature Palette provides examples of common industry components for design creation.

Display the Feature Palette. Click **Tools**, **Feature Palette** from the Main menu. Double-click the **component folder**, (SolidWorks\data\Palette Parts), Figure 2.22.

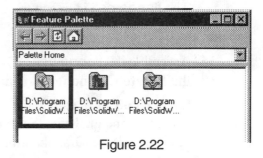

Figure 2.22

Double-click the **hardware folder** hardware icon. The hardware Feature Library components are displayed.

The flange bolt flange bolt icon represents a family of similar shaped components in various configurations, Figure 2.23.

Figure 2.23

Add a flange bolt to the assembly. Drag the **flange bolt** icon to left of the GUIDE-ROD assembly into the Graphics window. Release the **mouse button**.

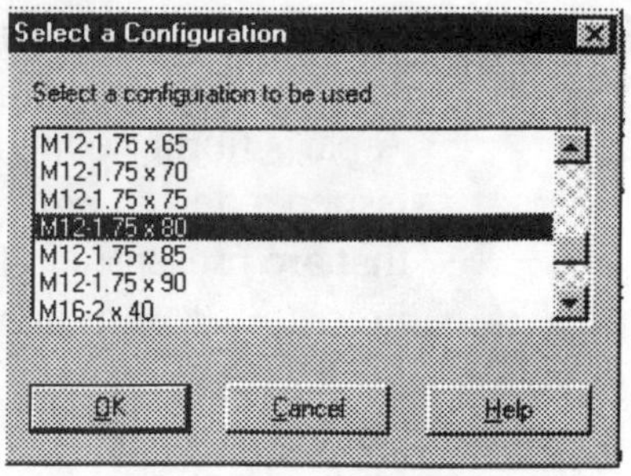

The 10 mm bolt contained in the Feature Palette is not long enough for the GUIDE.

Figure 2.24

Select a 12 mm flange bolt, **M12-1.75 x 80** from the configuration list, Figure 2.24. Click **OK**.

The flange bolt appears in the assembly, Figure 2.25. **Close** the Feature Palette.

The flange bolt is free to travel and rotate.

Assemble the flange bolt. Click the **Rotate Component Around Center Point** ⟲ icon in the Assembly menu. Click the **shaft** of the flange bolt. Drag the **mouse pointer** in a vertical direction. The flange bolt rotates, Figure 2.26a.

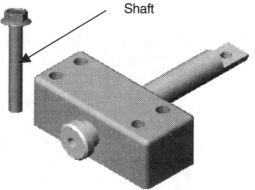

Figure 2.25

### 2.9.2   SmartMates

A SmartMate occurs when a component is placed into an assembly. The mouse pointer displays a SmartMate feedback symbol when common geometry and relationships exist between the component and the assembly.

SmartMates are Concentric or Coincident Planes. A Concentric SmartMate assumes that the geometry on the component has the same center as the geometry on an assembled reference. A Coincident Planes SmartMate assumes that a plane on the component lies along a plane on the assembly.

Shaft

Figure 2.26a

As you drag components into place, the mouse pointer provides feedback such as:

- Concentric

- Coincident

Place the first of the four **flange bolt**.

Create the first Mate. The cylindrical face of the **flange bolt** mates concentric with the cylindrical face of the top left hole.

Click the **SmartMate** icon in the Assembly toolbar. Double-click the **cylindrical face** of the **flange bolt**, Figure 2.27a. The mouse pointer displays .

Drag the **flange bolt** to the front left hole of the GUIDE.

The SmartMate mouse pointer displays Concentric . Do not release the mouse button. Press the **Tab** key if the **flange bolt** is upside down. Release the **mouse button**, Figure 2.27b.

Create the second mate. The flat bottom face of the **flange bolt** is coincident with the top face of the GUIDE.

The face of the **flange bolt** is difficult to view. Rotate the part to improve visibility.

Click the **Rotate** icon from the View menu. Drag the **mouse pointer** in a vertical direction. View the flat bottom face of the cylindrical flange, Figure 2.28.

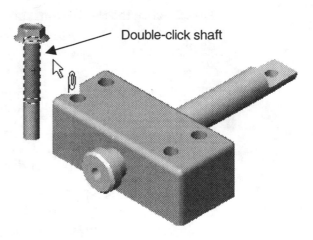

Double-click shaft

Figure 2.27a

Figure 2.27b

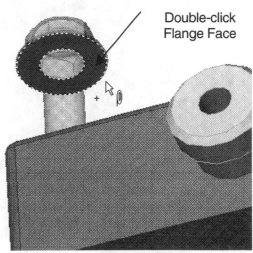

Double-click
Flange Face

Figure 2.28

Double-click the **flat bottom face** of the flange bolt. The mouse pointer displays  .

Click the **Isometric**  icon from the View menu. Drag the **flange bolt** to the top face of the GUIDE.

The SmartMate mouse pointer displays Coincident Planes  , Figure 2.29. The top mating face of the GUIDE is displayed in green. Release the **mouse button**.

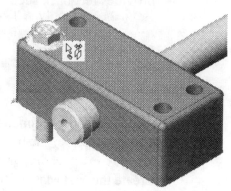

Figure 2.29

The flange bolt is under defined.

Create the third mate. The front face of the flange bolt head is parallel with the front face of the GUIDE. Click the **Mate**  icon from the Assembly menu. Click the **front face** of the hex head bolt. Click the **front face** of the GUIDE, Figure 2.30. Click **Parallel** from the Assembly Mating dialog box. Click **Apply**.

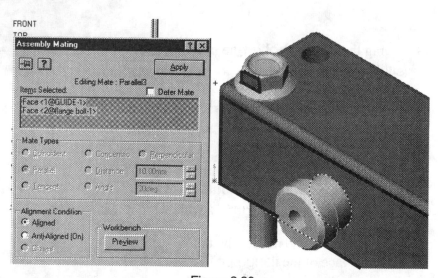

Figure 2.30

The flange bolt is fully defined.

The Mates and SmartMates are listed in the FeatureManager, Figure 2.31. Click *MateGroup1*.

Note: If you deleted a Mate and then recreate it, the Mate numbers will be in a different order.

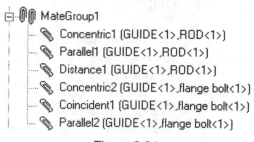

Figure 2.31

Example: Concentric3 (GUIDE<1>, flange bolt<1>) instead of the displayed Concentric2 (GUIDE<1>, flange bolt<1>).

**Save** the assembly. Click **Yes** to the question, "Save the document and referenced models now?" The referenced models are the GUIDE and ROD.

### 2.9.3   Modify the GUIDE

The GUIDE hole dimensions are too small for the bolts. Display the hole dimensions.

Double click the **GUIDE** in the FeatureManager. Double-click the *Holes* feature.

Click the hole diameter dimension, ⌀**10**. Enter **12**.

Note: 12 corresponds to the **flange bolt** diameter, Figure 2.32.

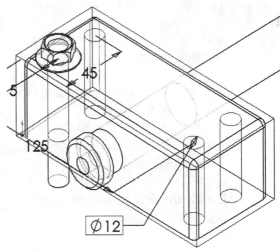

Figure 2.32

Update the GUIDE. Click the **Rebuild** ▮ icon. The hole dimensions are updated.

**Save** the Assembly.

### 2.9.4   Copy the Component

Copy components in an assembly.

Create the first **flange bolt**. Create three other **flange bolts** as copies. Each **flange bolt** requires a:

- Concentric SmartMate

- Coincident Plane SmartMate

- Parallel Mate

Create the second flange bolt.
Click the **flange bolt** name in
the FeatureManager,

⊞ 🔩 flange bolt<1> (M12-1.75 x 80)

Press the **Ctrl** key. Drag the
**flange bolt** name into the
Graphic Window. The mouse
pointer displays the component

icon, ⬚🔩 . Release the **mouse
button**, Figure 2.33.

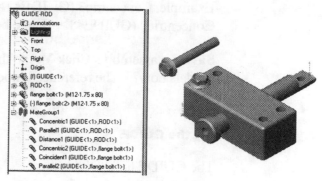

Figure 2.33

The instance number <2> appears after
the second flange bolt,
flange bolt<2> (M12-1.75 x 80)

in the FeatureManager.

Create the first SmartMate for the
second flange bolt.

Click the **SmartMate** 🔩 icon in the
Assembly toolbar. Double-click the
**cylindrical face** of the flange bolt. The
mouse pointer displays, ⬚∅ ,
Figure 2.34.

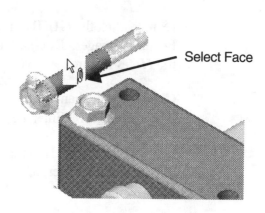

Select Face

Figure 2.34

Drag the second **flange bolt** to the back left GUIDE

hole. The mouse pointer displays Concentric ⬚∅ .
Do not release the mouse button. Press the **Tab** key
if the flange bolt is upside down. Release the **mouse
button**, Figure 2.35.

Create the second Mate. Rotate the assembly to view
the bottom flat face of the flange bolt. Click the
**Rotate** 🔄 icon.

Select the face geometry of the flange bolt. Click
the **Face Filter** 📇 icon located in the Filter toolbar.

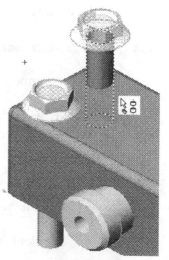

Double-click the **flat bottom face** of the flange bolt.

The mouse pointer displays ⬚∅ .

Figure 2.35

Click the **Isometric** icon from the View menu. Drag the **flange bolt** to the top face of the GUIDE. The mouse pointer displays the Coincident Planes, Figure 2.36. Release the **mouse button**.

Turn off the face filters. Click the **Face Filter** icon located in the Filter toolbar.

The flange bolt remains under defined.

Create the third Mate. Click the **Mate** icon in the Assembly menu. Click the **front face** of the hex head. Click the **front face** of the GUIDE, Figure 2.37. Click **Parallel** from the Assembly Mating dialog box. Click **Apply**. The flange bolt is fully defined.

Create the third flange bolt.

Copy the first **flange bolt**, flange bolt<1> (M12-1.75 x 80) . Press the **Ctrl** key down. Drag the **flange bolt** name into the Graphic Window. Create a **Concentric SmartMate**. Create a **Coincident Plane SmartMate**. Create a **Parallel Mate**.

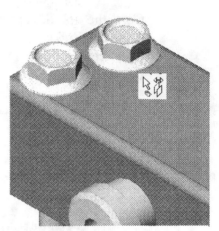

Figure 2.36

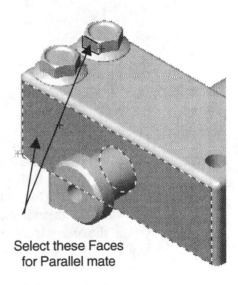

Select these Faces
for Parallel mate

Figure 2.37

Create the forth flange bolt.

Copy the first **flange bolt**.  Click the **flange bolt** 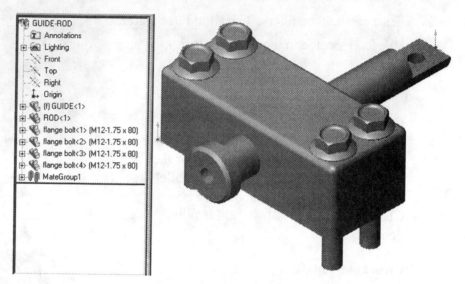 flange bolt<1> (M12-1.75 x 80) .
Press the **Ctrl** key.  Drag the **flange bolt** name into the Graphic Window.  **Create**
a **Concentric SmartMate**.  Create a **Coincident Plane SmartMate**.  Create a
**Parallel Mate**, Figure 2.38.

Figure 2.38

Display the created Mates in the FeatureManager.  Click the **Plus sign** ⊞ icon in
*MateGroup1*, Figure 2.39.

Note: There are two additional
keyboard commands that will assist
you in the *SmartMate* process: Shift
key and Alt key.

The Shift key is used to drag a
component into the assembly.  The
component is viewed with a normal
(opaque) appearance.  Release the
Shift key to display the transparent
appearance.

The Alt key is used to suspend
*SmartMates* while placing a
component into the assembly.

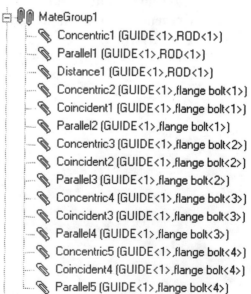

Figure 2.39

### 2.9.5   Modify Dimensions

The flange bolts are related to the GUIDE holes through the Mates constraints. If the location of the GUIDE holes change, the flange bolts move accordingly.

Expand the GUIDE component in the FeatureManager. Click the **Plus** ⊞icon. Select the *Holes* feature. Click **Top** from the Standard toolbar.

Double-click the horizontal dimension, **125**. Enter **115**. Click the **Apply** √ icon. Double-click **15**. Enter **25**, Figure 2.40. Click **Rebuild**.

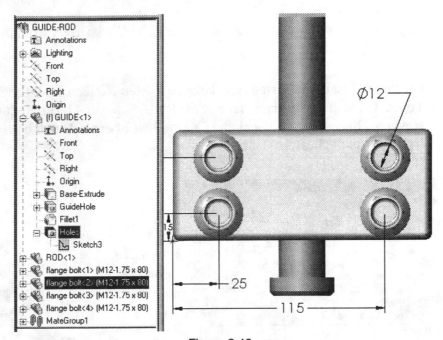

Figure 2.40

**Save** the assembly. Click **Yes** to the question, "Save the document and referenced models now?" The referenced models are the GUIDE and ROD.

## 2.10  Exploded View

Exploded views assist the designer in the viewing of the design creation.  You can fully explode an assembly in a single or multi step procedure.  AutoExplode explodes an assembly in a single step procedure.  In a multi step procedure, you create individual explode steps.

Display the *Isometric* view.  Click the **Isometric** ⬗ icon.

### 2.10.1  Explode Step 1

Explode the assembly in a multi step approach.  Click **Insert** from the Main menu.  Click **Exploded View**.

Create the first explode step.  Click the **New Step** ⬚ icon from the Step Editing Tools.  Click the **cylindrical face** of the ROD, Figure 2.41.  Enter **200**.  Check the **Reverse Direction** box.  Click the **Component to explode** text box.  Click **ROD<1>** in the FeatureManager.

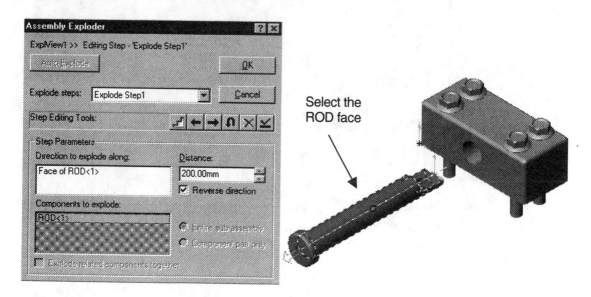

Figure 2.41

View the first exploded step, named Explode Step1.  Click the **Apply Check Mark** ⬓ icon.

### 2.10.2  Explode Step 2

Create the second step.  Click the **New Step** ⬛ icon.  Click the **vertical edge** of the GUIDE.  Enter **100**.  Click the **Reverse** check box, Figure 2.42.

Click **flange bolt <1>**.  Click **flange bolt <2>**.  Click **flange bolt <3>**.  Click **flange bolt <4>**.

Note: All 4 flange bolts are listed in the Component to explode list box.

View the second exploded step, named Explode Step2.  Click the **Apply Check Mark** ☑ icon.  Click **OK**.

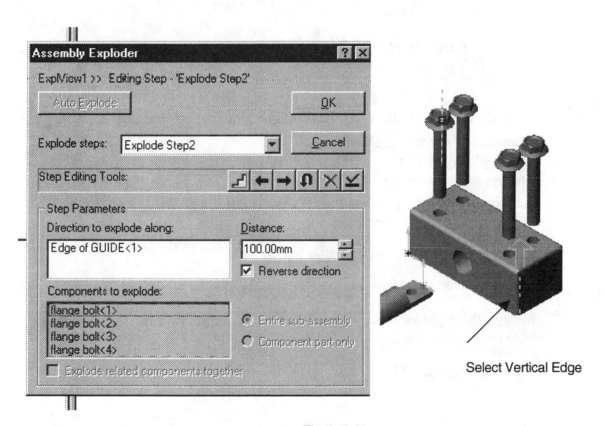

Figure 2.42

Fit the *Exploded* view to the Graphics window.  Press the **f** key.

### 2.10.3 Step Editing Tools and Viewing Exploded State

Edit steps in the *Exploded* view with the following Step Editing tools. From left to right the following icons represents:

- New Step

- Edit Previous Step

- Edit Next Step

- Undo Changes to Current Step

- Delete Current Step

- Apply the Current Step

View the exploded state. Click the **Configuration Manager** icon at the bottom of the FeatureManager. Expand the Default icon. Click the **Plus** icon to the left of the Default entry. Expand the ExpView1 icon. Click the **Plus** icon to the left of ExpView1 entry, Figure 2.43.

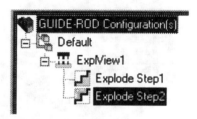

Figure 2.43

Return to the FeatureManager. Click the **FeatureManager** icon at the bottom of the Configuration Manager.

Remove the exploded state. Right-click in the **Graphic window**. Click **Collapse** from the Pop-up menu.

**Save** the Assembly.

## 2.11 Section View

*Section* views display the internal cross section of a component or assembly. The *Section* view dissects a model like a knife slicing through a stick of butter. *Section* views can be performed anywhere in a model. The location of the cut corresponds to the section plane.

A *Section* plane is a planar face or reference plane. In this section, you will construct a new reference plane. First, specify the type of plane to be constructed. Click **Insert**, **Reference Geometry**, **Plane** from the Main toolbar. Click **Offset** from the Specify Construction dialog box. Click **Next**, Figure 2.44.

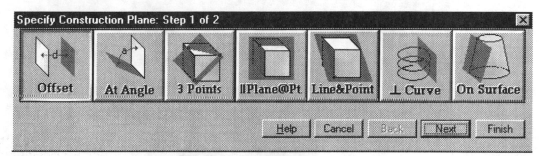

Figure 2.44

Next, specify the front face of the GUIDE as the Offset plane. Click the **front face** of the GUIDE from the Entity Selected text box. Enter **15** in the Distance spin box. Click the **Reverse Direction** check box. An Offset plane is drawn through the center of the front flange bolts, Figure 2.45.

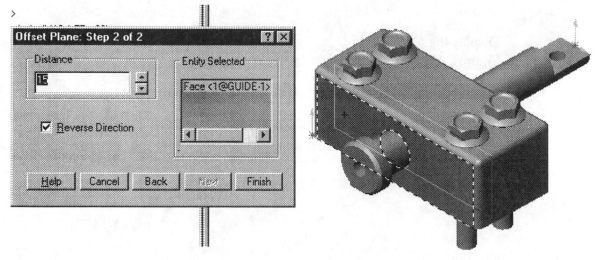

Figure 2.45

Create the Offset plane. Click **Finish**. The *Plane4* name appears in the FeatureManager. *Plane4* is displayed parallel to the front face of the GUIDE. Rename *Plane4* to *SectionPlane*.

Create the section view on *SectionPlane*. Click the **SectionPlane**. Click **View** from the Main menu. Click **Display**, **Section View**, Figure 2.46. Set the viewing direction of the *Section* cut. Click **Flip the Side to View** check box. Click **Preview** to display the section. Display the section in an *Isometric* view. Click the **Isometric** icon. Click **OK**.

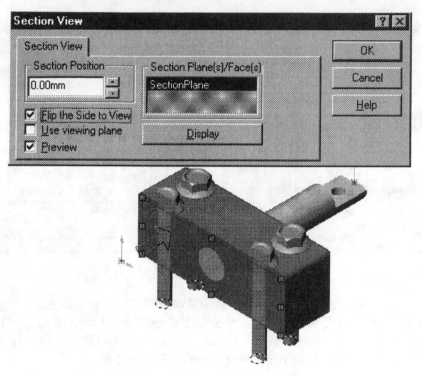

Figure 2.46

Display the *Front* view, Figure 2.47.

Click *Front* from the Standard Views toolbar.

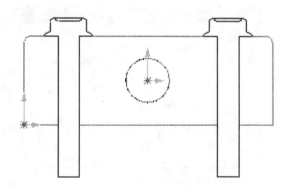

Figure 2.47

Display the complete view. Remove the check mark [✓ Section View]. Click the **View** from the Main menu. Click **Display**. Click **Section View**.

## 2.12 Suppress Features and Components

Suppressed features and components are not displayed. During model rebuilding, suppressed features and components are not calculated. This saves rebuilding time for complex models.

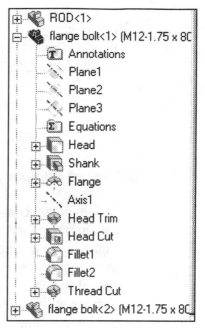

Figure 2.48

Features and components are suppressed at the component or assembly level in the FeatureManager. The names of the suppressed features and components are displayed in light gray.

In this section, modify the flange bolt to a hex head bolt. Update the change in the part. How do you do this? Answer: Suppress the *Flange* feature.

Expand **flange bolt<1>**. Double-click **flange bolt<1>** from the FeatureManager, Figure 2.48.

Suppress the *Flange* feature. Right-click on the *Flange* feature. Select **Properties** from the Pop-up menu, Figure 2.49.

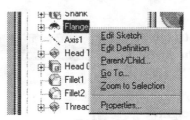

Figure 2.49

Click the **Suppressed** check box. Click **OK**, Figure 2.50.

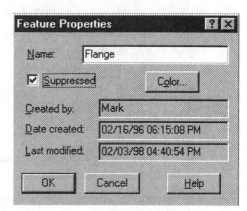

Figure 2.50

Hide all planes. Click **View** from the Main menu. Click **Planes**.

All bolts are updated to display
the suppressed *Flange* feature,
Figure 2.51.

**Save** the GUIDE-ROD. Click
**Yes** to the question, "Save the
document and referenced models
now?" The referenced models
are the parts, GUIDE and ROD.

Close all parts and assemblies.
Click **Window** from the Main
menu. Click **Close All**.

## 2.13 Lightweight Components

Figure 2.51

A lightweight component is a component that only loads a portion of its model
data into memory. A fully resolved component loads all model data into memory.
Lightweight components are more efficient for large assemblies. They are quicker
to open and Rebuild in an
assembly. After a lightweight
component is edited, the
component is fully resolved into
all of its features.

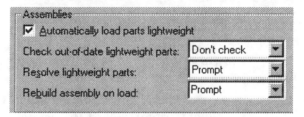

Figure 2.52

Load lightweight components.
Click **Tools** from the Main
menu. Click **Options**. Click the
**Performance** from the Systems Options tab. Click
the **Automatically load parts lightweight** check box
under the Assemblies box, Figure 2.52. Click **OK**.

Components are displayed in their lightweight
state when an assembly is opened, Figure 2.53. Edit
a component to remove its lightweight state.

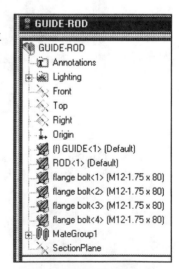

In the assembly process, you are required to
incorporate design changes to various components.
Changes are based on customer requirements,
materials, company standards, etc.

Figure 2.53

Work through additional exercises at the end of
Project 2 to practice Mates and SmartMates. A way of improving your ability is
to experiment with different Mates relationships. The first relationship may not be
optimum. Experiment with deleting and creating Mates.

## 2.14 Questions

1.  Describe an assembly or sub-assembly.

2.  What are Mates and why are they important in assembling components?

3.  Name and describe the three major types of Fits.

4.  Name and describe the two assembly modeling techniques in SolidWorks.

5.  Describe Dynamic motion.

6.  In an assembly, each component has_____# degrees of freedom.  Name them.

7.  True or False.  A Fixed component cannot move and is locked to the *Origin*.

8.  Describe the function of Selection Filters.

9.  What is the SolidWorks Feature Palette?

10. Describe the different types of SmartMates.

11. How are SmartMates used?

12. Describe a Section View.

13. What is a lightweight component?  Describe the features and benefits.

14. What are *Suppressed* features and components?  Provided an example.

## 2.15 Exercises

### I. Create the Assemblies.

Exercise 2.1

a) Create two parts:

- L-BRACKET

- PLATE

The Base component is the PLATE.

Exercise 2.1

b) Review the following assembly
commands before mating the L-BRACKET:

- Move component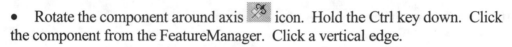

- Rotate the component around axis 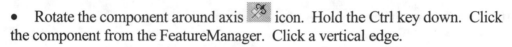 icon.  Hold the Ctrl key down.  Click the component from the FeatureManager.  Click a vertical edge.

- Rotate component around center point

c) Complete the assembly.  Utilize one PLATE and two L-BRACKETs and two BOLTS.

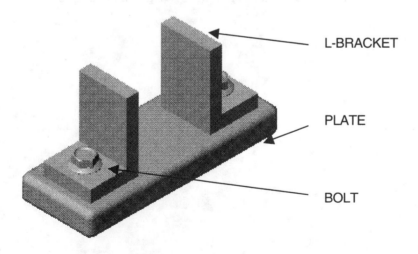

L-BRACKET

PLATE

BOLT

Exercise 2.2          Create the SUPPORT assembly from the PLATE and STANDOFF.

Exercise 2.3          Create the HANGER assembly from the parts, U-BRACKET and PIN.

Exercise 2.4          Create the FIXTURE assembly from the parts, MOUNTING PLATE and T-SECTION.

SUPPORT assembly          HANGER assembly          FIXTURE assembly

Exercise 2.2          Exercise 2.3          Exercise 2.4

Notes

# Project 3

## Fundamentals of Drawing

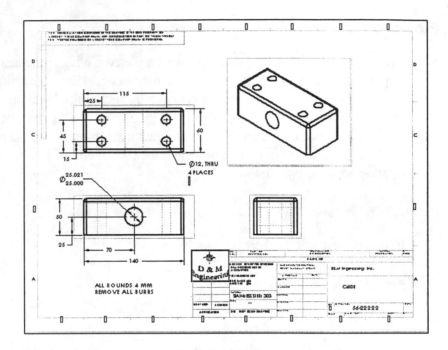

Below are the desired outcomes and usage competencies based upon completion
of Project 3.

| Project Desired Outcomes | Usage Competencies |
|---|---|
| • A GUIDE drawing with a customized sheet format containing a company logo, title block and sheet information. | • Ability to create a custom sheet format, logo, title block and sheet information. |
| • A GUIDE-ROD drawing with a Bill of Materials. | • Knowledge to develop and incorporate a Bill of Materials into a drawing. |

## 3  Project 3 – Fundamentals of Drawing

### 3.1  Project Objective

Create a GUIDE drawing.  Create a GUIDE-ROD drawing.

### 3.2  Project Situation

The individual parts and assembly are completed.  What is the next step?  You are required to create drawings for various internal departments, namely: production, purchasing, engineering, inspection and manufacturing.  Each drawing may contain unique information and specific footnotes.  Each department requires distinct information.  Example: A manufacturing drawing would require information on assembly, Bill of Materials, fabrication techniques and references to other relative information.

### 3.3  Project Overview

In this project you will create two drawings:

- GUIDE

- GUIDE-ROD

The GUIDE drawing contains three standard views, (principle views) and an *Isometric* view.  Do you remember what the three standard views are?  They are: *Top*, *Front* and *Right* views, Figure 3.1.  Orient the views to fit the drawing sheet.  Incorporate the GUIDE dimensions into the drawing.

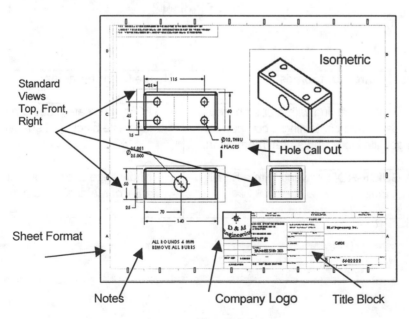

Figure 3.1

The GUIDE-ROD assembly drawing contains an exploded view. The drawing contains a Bill of Materials and balloon text, Figure 3.2.

Note: Microsoft EXCEL 97 or a later version is required to create the Bill of Materials.

Both drawings utilize a custom sheet format containing a company logo, title block and sheet information.

There are two major design modes used to develop a drawing: Edit Sheet Format and Edit Sheet.

The Edit Sheet Format mode provides the ability to:

- Change the title block size and text headings

- Incorporate a logo

- Add drawing, design or company text

The Edit Sheet mode provides the ability to:

- Add or modify views

- Add or modify dimensions

- Add or modify text

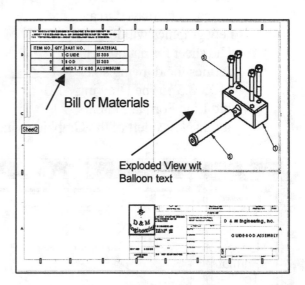

Figure 3.2

### 3.4  Create a Drawing Sheet Format

A drawing sheet format is the foundation for drawing information. A drawing sheet format can specify: drawing size, company information, manufacturing / assembly requirements and more.

There are various sheet format options in SolidWorks:

- Standard SolidWorks Sheet Format

- Custom Sheet Format

- No Sheet Format

In this section, create a custom sheet format. First, start with the standard SolidWorks sheet format.

Create the GUIDE drawing.
Click the **New** ☐ icon. Click
**Drawing**, Figure 3.3. Click
**OK**.

Figure 3.3

The Sheet Format To Use
dialog box is displayed.
Select **A-Landscape** from
the Standard Sheet Format
drop down list, Figure 3.4.
Click **OK**.

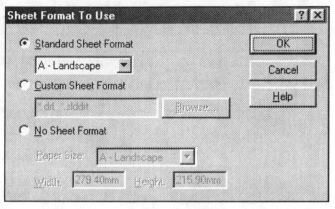

Figure 3.4

The A-Landscape sheet
format is displayed in a
new graphics window.
The sheet format border
defines the drawing size,
11" x 8.5". The Drawing
and Line Format toolbars
are displayed left of the Graphics window, Figure 3.5.

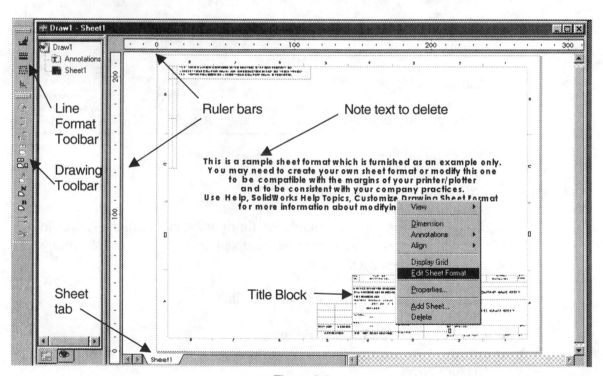

Figure 3.5

Use the sliding ruler bars to position views and text on the drawing sheet. The
Sheet1 tab defines the active drawing sheet. A drawing may consist of multiple
sheets. Right-click in the **Graphics window**. Click **Edit Sheet Format** from the
Pop-up menu. The drawing title block lines turn blue.

Click the **Note Text** in the center of the drawing. The note text boundary displays 4 green squares, Figure 3.6. Delete the text. Press the **Delete** key.

Note: You cannot modify Sheet Format text, lines or title block information unless you are in Edit Sheet Format mode.

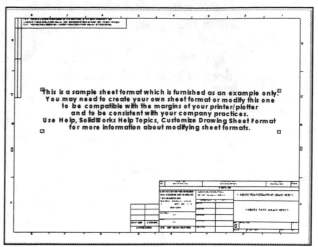

Figure 3.6

### 3.4.1   Title Block

The title block contains vital part or assembly information. Each company can have a unique version of a title block. Most title blocks contain the following type of information:

- Company Name/Logo
- Part name
- Drawing description
- Sheet number
- Tolerance
- Sheet size
- CAD file name
- Quantity required
- Checked by

- Part number
- Drawing number
- Revision number
- Material & finish
- Drawing scale
- Revision block
- Engineering Change Orders
- Drawn by
- Approved by

The title block is located in the lower right hand corner, Figure 3.7.

View the right side of the title block. Click the **Zoom to Area** on the sheet format title block.

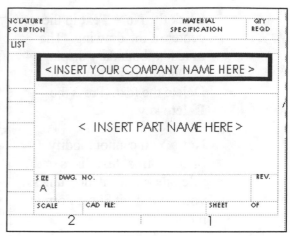

Figure 3.7

Right-click on the <**INSERT YOUR COMPANY NAME HERE**> text. Click **Properties**, Figure 3.8a.

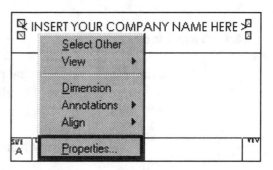

Figure 3.8a

The Properties dialog box appears, Figure 3.8b.

Enter a company name. Click inside the **Note text box**.

Example: Enter **D & M Engineering, Inc**.

Note: You can specify the following: text justification, rotation angle, font size and type from the Properties toolbar.

Uncheck the **Use document font** check box and click the **Font** button to change the font size.

Accept the text. Click **OK**.

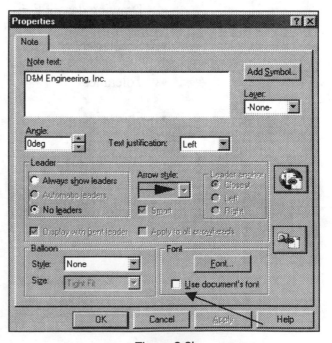

Figure 3.8b

The text is displayed in the title block, Figure 3.8c.

Figure 3.8c

### 3.4.2   Save the Sheet Format

Save the sheet format as a custom sheet format.

Click **File**, **Save Sheet Format**, Figure 3.9.

The Save Sheet Format dialog box appears, Figure 3.10.  Click the **Custom Sheet Format** button.

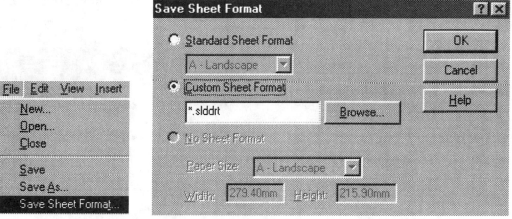

Figure 3.9

Figure 3.10

Click the **Browse** button from the Save Sheet Format dialog box.

The default Sheet Format folder is called data, Figure 3.11.  The file extension for Sheet Format is .slddrt.

Enter File name.  Enter **CUSTOM-A**.  Click **Save** from the Save Sheet Format dialog box.

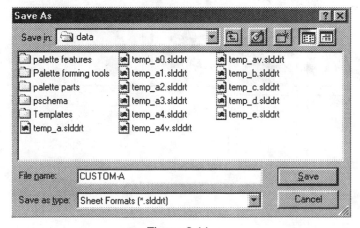

Figure 3.11

### 3.4.3   Company Logo

A company logo is normally located in the title block.

Create a company logo by copying a picture file from Microsoft ClipArt using Microsoft Word.  Copy / Paste the logo into the SolidWorks drawing.

Note: The following logo example was created in Microsoft Word 2000 using the COMPASS.wmf and WordArt text.  Any ClipArt picture, scanned image or bitmap can be used.

Create a New Microsoft Word Document.  Click the **New** ⬚ icon from the Standard toolbar in MS Word.  Click the **ClipArt** 🖼 icon from the Draw toolbar, Figure 3.12.

Rectangle          WordArt          ClipArt

Figure 3.12

Double-click on the COMPASS.wmf file.  Insert the **compass.wmf** picture file

from the ClipArt menu, Figure 3.13a.

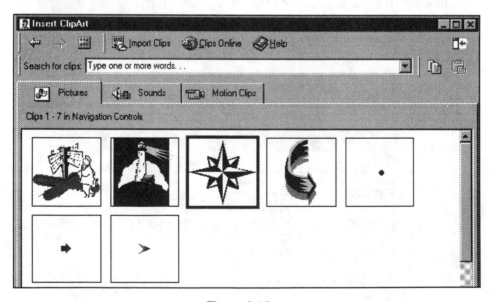

Figure 3.13a

Next, add text to the logo picture. Click the **Insert Word Art** icon from the Draw toolbar. Click a **WordArt** style, Figure 3.13b.

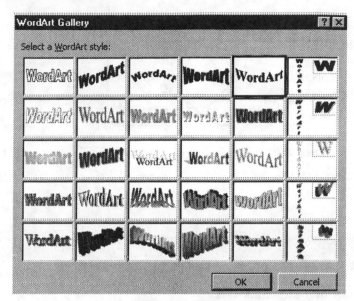

Figure 3.13b

Enter **D & M Engineering** in the text box.

Click **24** from the Size drop down list. Click **OK**, Figure 3.13c.

Drag the **Word Art text** under the picture. Size the **Word Art text** by dragging the picture handles, Figure 3.13d.

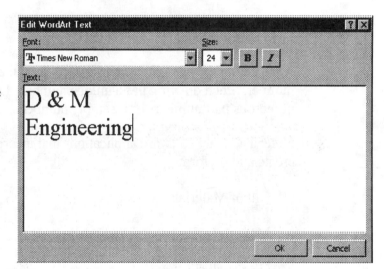

Figure 3.13c

Select the geometry.
Press the **Shift** key. Select the **compass picture**.

Select the **WordArt text**. Click the **Copy** icon.

The logo is placed into the Clipboard. Close Microsoft Word. Click **File, Exit**.

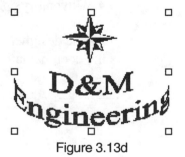

Figure 3.13d

Place the logo into the title block. Click inside the SolidWorks **Graphics window**, on the left-hand corner of the title block.

Click the **Paste** icon. Size the **logo** to the SolidWorks title block by dragging the picture handles, Figure 3.14a and Figure 3.14b.

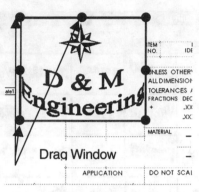

Figure 3.14a

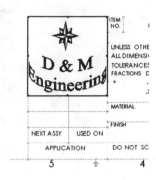

Figure 3.14b

Fit the drawing to the Graphics window. Press the **f** key.

Update the custom sheet format. Click **File, Save Sheet Format**. Click **Custom Sheet Format**. Click **OK**.

### 3.4.4 Part Number and Part Name

A part number is a numeric representation of the part. Each part has a unique number. Each drawing has a unique number. Drawings can incorporate numerous part numbers or assembly numbers.

Note: There are software applications that incorporate unique part numbers to create and perform:

- Bill of Materials

- Manufacturing procedures

- Cost analysis

- Inventory control / Just in Time, JIT

As the designer, you are required to procure the part and drawing numbers from the documentation control manager.

In this project, use the following prefix codes to categories created parts and drawings. The part name, part number and drawing numbers are as follows:

| Category | Prefix | Part Name | Part Number | Drawing Number |
|----------|--------|-----------|-------------|----------------|
| Machined Parts | 56- | GUIDE | 56-A26 | 56-22222 |
|  |  | ROD | 56-A27 | 56-22223 |
| Purchased Parts | 99- | FLANGE BOLT | 99-FBM12x80 | 999-551-12-80 |
| Assemblies | 10- | GUIDE-ROD | 10-A123 | 10-50123 |

Enter part name. Double-click on the <**INSERT PART NAME HERE**> text, Figure 3.15a.

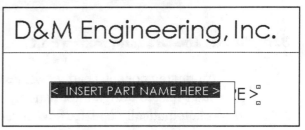

Figure 3.15a

Enter **GUIDE** in the Note Text box. Click **OK**, Figure 3.15b.

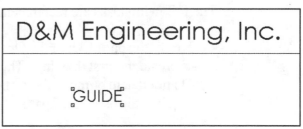

Figure 3.15b

Reposition the text. Click and Drag the **GUIDE text** to the left, Figure 3.15c.

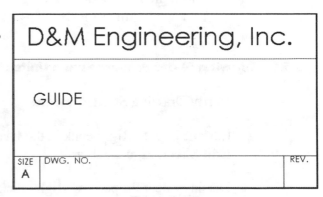

Figure 3.15c

### 3.4.5   Save the Drawing

Each drawing has a unique file name.  In SolidWorks, drawing file names end with a .slddrw suffix.  Part file names end with a .sldprt suffix.

A drawing or part file can have the same prefix.  A drawing or part file cannot have the same suffix.  Example: Drawing file name: GUIDE.slddrw.  Part file name: GUIDE.sldprt.

Save the drawing.  Click **File**, **Save As** from the Main menu.  Enter the drawing name.  Enter **GUIDE**, Figure 3.16.  Click **Save**.

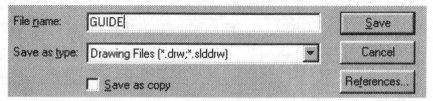

Figure 3.16

## 3.5   Create the Drawing from a Part

A drawing contains part views, geometric dimensioning and tolerances, notes and other related design information.  In SolidWorks, if a part is modified, the drawing is automatically updated.  When a dimension in the drawing is modified, the part is automatically updated.

The following tasks are recommended before starting the GUIDE:

*   Verify the part: Click File, Open.  Select GUIDE.  Note: This is the part used to create the first drawing.  The part shares the same database as the drawing.  Do not delete or move the part.  The drawing will not be valid.  The drawing requires the associated part.

*   View dimensions in each part.  From the FeatureManager, slide the Rollback bar to the *Extruded-Base* feature.  Step through each feature of the part and review all dimensions.

### 3.5.1   Drawing Sheet Properties and Units

Verify Drawing Sheet Properties.

Edit the Sheet.  Right-click in the **Graphics Window**.  Click **Edit Sheet**, Figure 3.17a.

Display Sheet Properties.  Right-click in the **Graphics window**.  Click **Properties**.

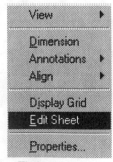

Figure 3.17a

The Sheet Setup dialog box is displayed, Figure 3.17b.

Set the Sheet Scale to a 1:2 ratio. Enter **2** in the second Scale text box.

The orientation of the three views: *Top*, *Front* and *Right* are based on third angle Orthographic project. Set the Type of projection to Third angle. Click the **Third angle** button.

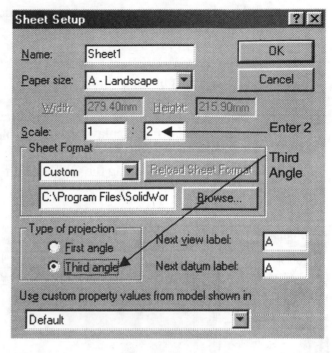

Figure 3.17b

Insure that the Drawing Sheet Units are set to Millimeters. Click **Tools** from the Main menu. Click **Options**. Click the **Document Properties** tab. Click **Units**. Select **Millimeters** from the Linear Units drop down list. Click **OK**.

### 3.5.2 Add Standard Views

Add three standard views to the drawing: *Top*, *Front* and *Right*.

**Close** all other windows, except for GUIDE.slddrw.

Open the GUIDE part. Click the **Open** icon. Enter **GUIDE.sldprt**.

Display the drawing. Click **Window** from the Main menu, Figure 3.17b. Click **GUIDE-Sheet1**.

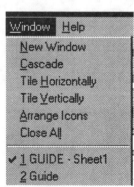

Figure 3.17b

The drawing must be in Edit Sheet mode in order to insert the drawing views. Click **Tile Horizontal**, Figure 3.18.

Drag the **GUIDE Part** icon from the FeatureManager into the center of the drawing graphics window.

Note: You cannot insert a part into a drawing when Edit Sheet Format is selected. Edit Sheet Format displays all lines in blue. The 3 drawings views will flash on the screen once and then disappear. Right-click in the Drawing Graphics window. Click Edit Sheet. Drag the GUIDE Part icon into the Drawing Sheet.

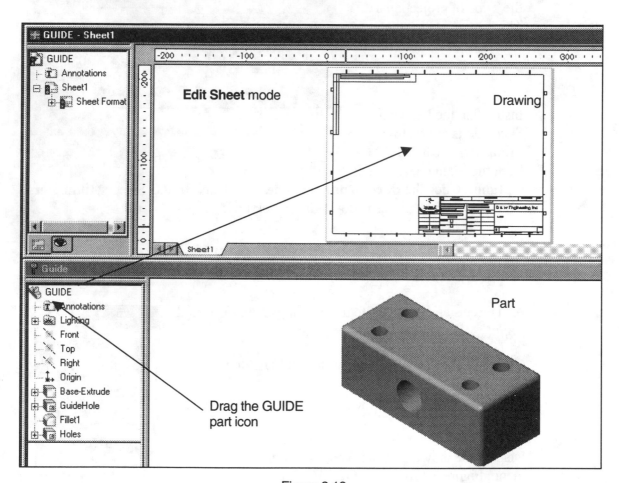

Figure 3.18

The three standard views are displayed in the graphic window. Click the **Maximize** icon for the GUIDE drawing window, Figure 3.19.

Note: It may take a few seconds to create the standard views depending upon your computer configuration.

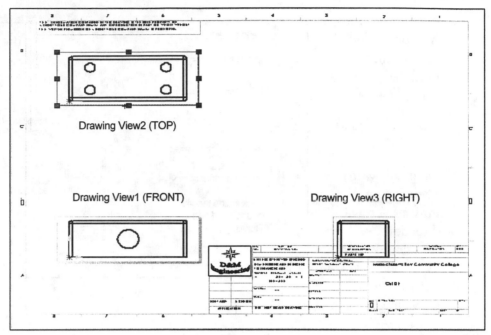

Figure 3.19

The mouse pointer provides feedback in both the Drawing Sheet and Drawing View modes. The mouse pointer displays the Drawing Sheet icon when the Sheet properties and commands are executed. The mouse pointer displays the Drawing View icon when the View properties and commands are executed.

### 3.5.3 Move Views

Reposition views on a drawing. Move the views. Click the view boundary of *Drawing View1* (Front). The mouse pointer displays the Drawing View icon. The view boundary is displayed in green, Figure 3.21.

Position the **mouse pointer** on the edge of the view until the Drawing Move View icon is displayed.

Drag *Drawing View1* in an upward vertical direction. *Drawing View2* (Top) and *Drawing View3* (Right) move aligned to *Drawing View1* (Front).

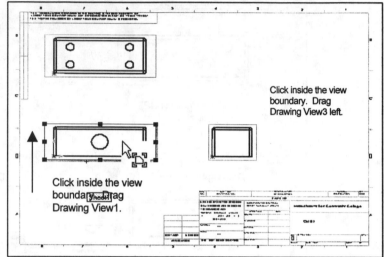

Click inside the view boundary. Drag Drawing View3 left.

Click inside the view boundary. Drag Drawing View1.

Figure 3.21

Move *Drawing View3* in a right to left direction. Click the ***Drawing View3*** view boundary. Position the **mouse pointer** on the edge of the view until the Drawing Move View icon is displayed.

Drag ***Drawing View3*** in a right to left direction towards *Drawing View1*.

Move *Drawing View2* in a downward vertical direction. Click the **view border**. Drag ***Drawing View2*** in a downward direction towards *Drawing View1*.

Provide approximately 1" to 2" (25 – 50 mm) between each view for dimension placement.

### 3.5.4 Add a Named View

Named views display the part or assembly in various orientations. Add named views to the drawing at anytime.

Add an *Isometric* view to the drawing. Click the **Named View** icon from the Drawing toolbar. The mouse pointer displays the Select Model icon.

Click inside ***Drawing View1***(Front).

The named views
for the GUIDE are
displayed,
Figure 3.22.  Click
*Isometric*, **OK**
from the Drawing
View – Named
View dialog box.

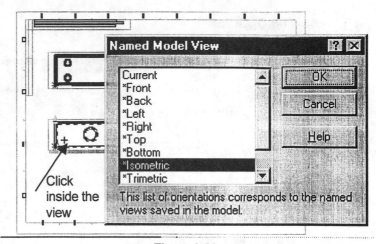

Figure 3.22

Click **Window**
from the Main
menu.  Click
**GUIDE**.  The
*Isometric* view is
placed on the mouse pointer.  Position the *Isometric* view.  Click the **Graphics window** to the right of *Drawing View2*, Figure 3.23.

Click **Yes** to the question: Do you want to switch the view to use Isometric (True) dimensions?

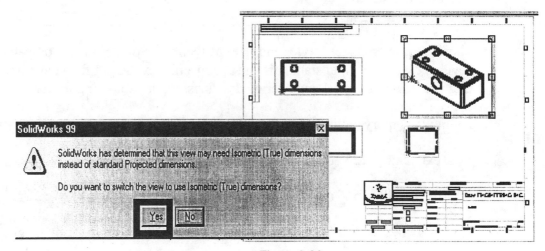

Figure 3.23

**Save** the GUIDE drawing.

## 3.6    Detailing Options

The Detailing options provide the ability to address: dimensioning standards, text style, center marks, witness lines, arrow styles, tolerance and precision.

Display the Detailing options dialog box. Click **Tools** from the Main menu. Click **Options**. Click the **Document Properties** tab. The Detailing entry is the default, Figure 3.24a.

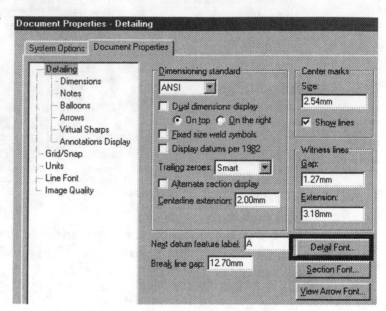

Figure 3.24a

There are numerous text styles and sizes available in SolidWorks. Companies develop drawing format standards and use specific text height for Metric and English drawings. Most engineering drawings use the following format:

- Font: Century Gothic – All capital letters.

- Text height: .125" or 3.5 mm for drawings up to B Size, 17" x 22".

- Text height: .156" or 5 mm for drawings larger than B Size, 17" x 22".

- Arrow heads: Solid filled with a 1:3 ratio of arrow width to arrow height.

Change the default text height. Click the **Detail Font** button from the Dimensions box. The Choose Font dialog box is displayed, Figure 3.25.

Set the dimension text. Click **Century Gothic** from the Font list box. Click the **Units** button. Enter **3.5** for height. Click **OK**.

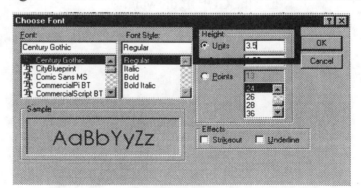

Figure 3.25

Change the arrow height.  Click the **Arrows** entry on the left side, below the Detailing entry.  The Detailing - Arrows dialog box is displayed, Figure 3.26.

Enter **1** for arrow Height.  Enter **3** for arrow Width.  Enter **6** for arrow Length.

Set the arrow style.  Click the solid **filled arrow head** from the Edge/vertex list box.

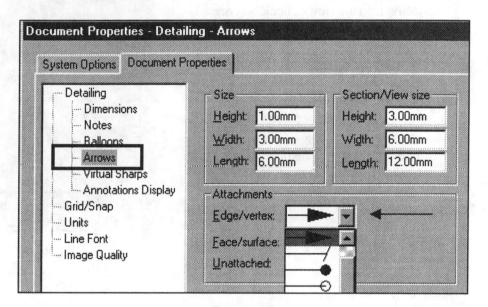

Figure 3.26

Click **OK** from the Document Properties dialog box.

**Save** the GUIDE drawing.

## 3.7  Insert Dimensions from the Part

Insert part dimensions into the drawing.

Click **Insert** from the Main menu.  Click **Model Items**.  The Insert Model Items dialog box is displayed, Figure 3.27. The Dimension check box and Import items into all views check box are checked.

Display all dimensions.  Click **OK**, Figure 3.28.

Note: Display dimensions in one drawing view.  Click *Drawing View2*. The view name is displayed in the Import Into Views list box.

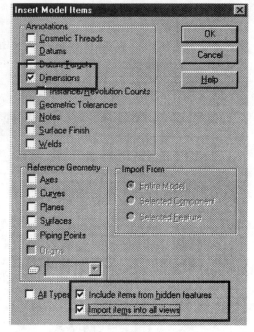

Figure 3.27

The dimensions are located too far from the profile lines.  You will move them later in this section.

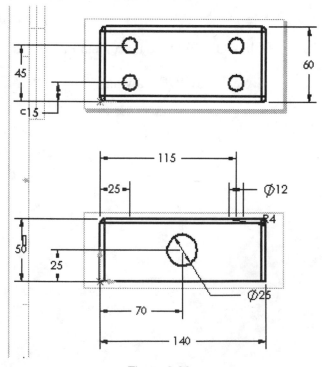

Figure 3.28

Drawing dimension location is dependent upon:

- How they were created in the part

- The order that the drawing views are selected

### 3.7.1   Move Dimensions in the Same View

Move dimensions within the same view.  Move linear dimensions in *Drawing View1*.  Click the vertical dimension text **25**, Figure 3.29a.  A green box is drawn around the text, Figure 3.29b.

Drawing View1

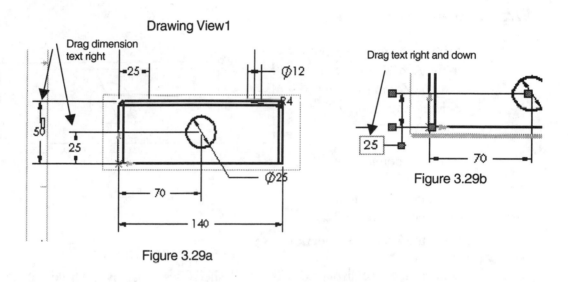

Figure 3.29a

Figure 3.29b

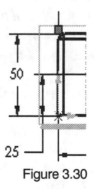

Figure 3.30

The mouse pointer displays the Linear Dimension  icon.  Drag the **dimension text** to the right.  Drag the **text** downward and position it outside of the arrows.

Click the vertical dimension text **50**.  Drag the **text** to the right, Figure 3.30.

Move the linear
dimensions in *Drawing
View2*. Click the vertical
dimension text **15**. Drag
the text to the right.

Click the vertical
dimension text **45**. Drag
the text to the right,
Figure 3.31a.

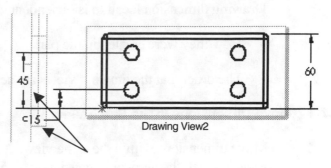

Figure 3.31a

### 3.7.2   View Display in Drawing Views

Modify the view display
in the *Drawing* views.
Use the Zoom to
Selection function to
enlarge the view.

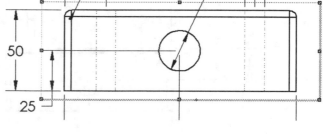

Click the boundary of
***Drawing View1***. The
green boundary line is
displayed, Figure 3.32.

Click the **Zoom to Selection** icon.

Figure 3.32

Hidden lines are displayed with thin dashed lines. Display the hidden lines. Click
the **Hidden In Gray** icon from the View toolbar. Fit the drawing to the
Graphics window. Press the **f** key. Click inside ***Drawing View 2***. Click the
**Hidden In Gray** icon. Repeat for ***Drawing View 3***.

### 3.7.3   Move Dimensions to a Different View

Move the dimensions from *Drawing
View1* to *Drawing View2*. Move the
linear dimensions 25 and 115 and the
hole diameter Ø12, Figure 3.33.

Note: The leader line points from the
dimension text to the profile of the
circle. The hole is displayed as two
parallel hidden lines in *Drawing
View1*. The hole is displayed as a
circle in *Drawing View2*.

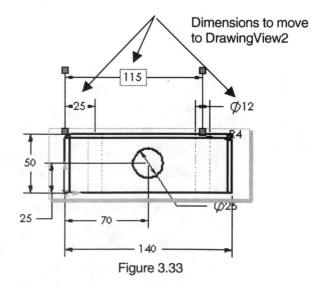

Figure 3.33

Hold the **Shift** key down. Click the horizontal dimension **25**. Drag the **dimension text** from *Drawing View1* to *Drawing View2*. Release the **mouse button** and the **Shift** key when the pointer is inside the *Drawing View2* boundary, Figure 3.34.

Drag dimension **25** above *Drawing View2*.

Note: When moving dimensions from one view to another, only drag the dimension text. Do not drag the leader lines. The text will not switch views if you drop it outside of the view boundary.

Hold the **Shift key** down. Click **115**. Drag **115** from *Drawing View1* to *Drawing View2*. Release the **mouse button** and the **Shift** key when the pointer is inside the *DrawingView2* boundary.

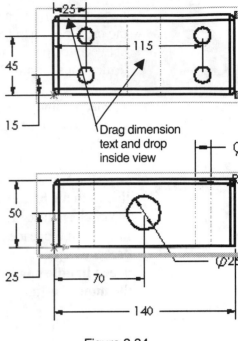

Figure 3.34

Drag **115** above *Drawing View2*, Figure 3.35.

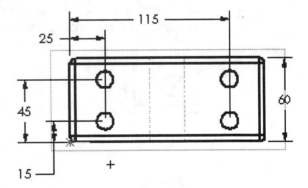

Figure 3.35

### 3.7.4   Edit Leader Lines

Leader lines reference the size of the profile. A gap must exist between the profile lines and the leader lines, Figure 3.36.

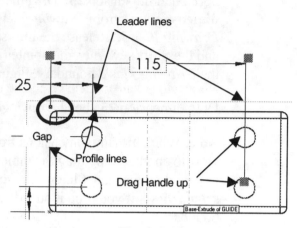

Figure 3.36

The leader lines must be shortened to maintain a drawing standard.

Drag the **end point** of the leader line from the bottom circle to the top circle. Click **115**. The handles of the leader lines turn green. Drag the **lower right green handle** upward from the bottom right circle to the top right circle. Release the **mouse button**.

Click **25**. The handles of the leader lines turn green. Drag the **lower right green handle** upwards from the bottom left circle to the top left circle. Release the **mouse button**.

## 3.8   Dimension Holes

Holes and other circular geometry can be dimensioned in three ways: Diameter, Radius and Linear (between two straight lines), Figure 3.37.

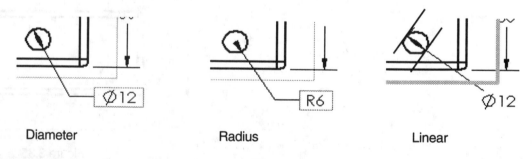

Diameter                    Radius                    Linear

Figure 3.37

The holes in *Drawing View2* require a diameter dimension and a note to represent all 4 holes. The dimension is currently displayed in *Drawing View1* as two pairs of hidden lines with one linear dimension, Ø12.

Hide the diameter dimension. Right-click on the dimension, ∅**12**. Click **Hide**, Figure 3.38. The dimension is not displayed.

Note: Click View from the Main menu. Click Hide/Show Dimensions to restore the dimension in the drawing view. Dimensions that are deleted can be reinserted into the drawing by selecting Insert, Model Items from the Main menu.

All four holes in Drawing View 2 and the large hole in Drawing View 1 require center marks and hole callouts.

Figure 3.38

### 3.8.1   Create Center Marks

Hole centerlines are composed of alternating long and short dash lines. The lines identify the center of a circle, axes or cylindrical geometry. Center Marks represents two perpendicular intersecting centerlines.

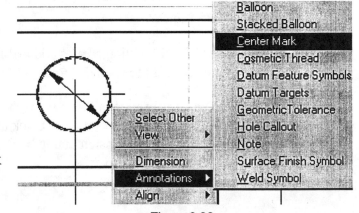

Place center marks on the hole in *Drawing View1* (Front). Activate the Center Mark. Right-click in the **Graphics window**. Click **Annotations**. Click **Center Mark**, Figure 3.39.

Figure 3.39

The mouse pointer displays the Center Mark icon. Click the **circumference** of the hole, Figure 3.40.

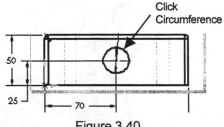

Place center marks on the 4 holes in *Drawing View2* (Top). Click the **circumference** of each hole, Figure 3.41.

Figure 3.40

Deactivate the Center Mark. Right-click in the **Graphics window**. Click **Annotations**. Click the **Center Mark**.

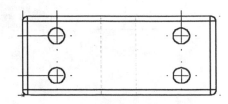

Figure 3.41

The Center Mark lines are too large.
Modify the size of the Center Mark lines.
Position the **mouse pointer** on the Center
Mark in the lower right circle. The mouse
pointer displays the center mark ⊕ icon.

A gap between the lines is required. Right-
click on the **Center Mark** in *Drawing
View2*. Click **Properties**. The Center
Mark dialog box appears, Figure 3.41.

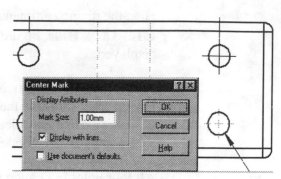
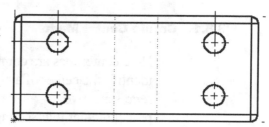

Figure 3.41

Change the size of the Center Mark.
Uncheck the **Use documents defaults** check
box. Enter **1.00** in the Mark Size box. Click
**OK**. Change the center mark for all holes in
*Drawing View 2*, Figure 3.42.

### 3.8.2   Create a Hole Callout

The Hole Callout function creates additional
notes required to dimension the holes. Create a Hole Callout.

Figure 3.42

Click the **circumference** of the lower right circle. Activate the Hole Callout.
Right-click and click **Annotations**. Click the **Hole Callout**, Figure 3.43a. The
Modify Text of the Dimension dialog box appears, Figure 3.43b.

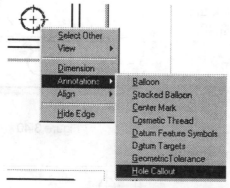

Figure 3.43a

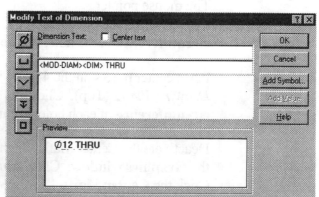

Figure 3.43b

Symbols are located on the left side of the Modify Text Dimension dialog box, Figure 3.44. The current text, |<MOD-DIAM><DIM> THRU is displayed in the text box.

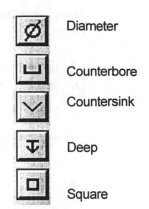

- <MOD-DIAM>: Diameter symbol ∅

- <DIM>: Dimension value 12

- THRU: Text determined by the original CUT-EXTRUSION for the holes

Represent all 4 holes with a single diameter dimension and note: 4 PLACES. Enter **4 PLACES** in the large text box, Figure 3.45. Display the text. Click **OK**, Figure 3.46.

Note: The mouse pointer displays the Hole Callout ⊔∅ icon, when the Hole Callout function is active. You will complete all Hole Callouts and then reposition the text.

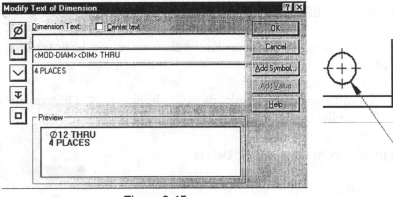

Figure 3.45

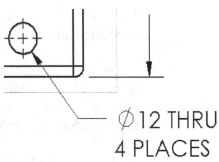

Figure 3.46

There are other ways to represent multiple holes with the same diameter. Place the number of holes (4) and the multiplication sign (X) before the diameter dimension. Position text on the same line as the dimension. Example:

4 X|<MOD-DIAM><DIM> THRU is displayed on the drawing as: 4 X ∅12.

Fit the drawing to the Graphics window. Press the **f** key.

The Hole Callout is activated. Click the **Edge** of the 25 mm circle in *Drawing View1* (Front).

The dimension text Ø25 THRU is displayed in the Preview window, Figure 3.47. Display the text. Click **OK**.

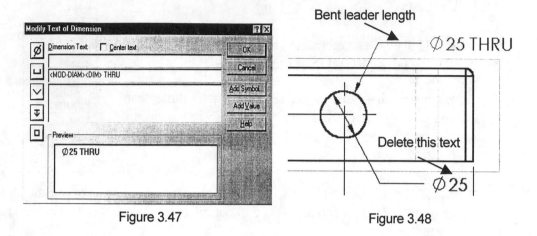

Figure 3.47                                                                 Figure 3.48

The Hole Callout modified the leader line arrow to the outside of the circle.

Deactivate the Hole Callout function. Right-click and click **Annotations**. Click the **Hole Callout**.

There are two dimensions for the hole diameter, Figure 3.48. Only one dimension is required. Delete the original dimension.

Click Ø**25** in the *Front* view. Click **Delete**.

Update the screen and display the new Hole Callout.

Repaint the screen. Click the **Repaint** icon from the Main toolbar.

Figure 3.49a

Note: If a Dimension Annotation function is activated, you cannot select the text, Figure 3.49a. To select text, deactivate the Dimension Annotation.

Center Marks, Hole Callouts and other Dimension Annotations are accessed through the Annotations toolbar menu. Right-click in the toolbar region. Click More Toolbars. Click Annotations, Figure 3.49b.

Figure 3.49b

The Bent leader line in the Hole Callout dimension is too long. Modify the Bent leader line length. Click **Tools** from the Main menu. Click **Options**. Click **Dimensions** from the Document Properties tab.

The Detailing – Dimensions dialog box is displayed, Figure 3.49c.

Enter **6.0** for the Bent leader line length.  Click **OK**, Figure 3.49d.

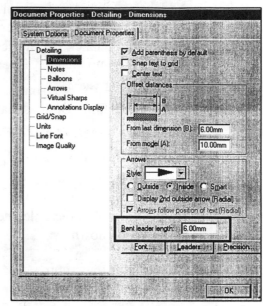

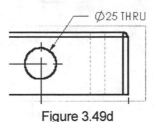

Figure 3.49d

Figure 3.49c

### 3.8.3    Clearance Fit for the ROD and GUIDE

In Project 2, you were introduced to the concept of tolerance and fit.  The design requires the ROD to slide through the GUIDE.  The shaft of the ROD is currently 25mm.  The Hole of the GUIDE is 25mm.  This is a problem!  You need to create a Clearance fit for the ROD and GUIDE.

Do you remember what a Clearance fit is?  A Clearance fit occurs when the shaft diameter of the ROD is smaller than the hole diameter of the GUIDE.  There are many different classes of clearance fits.  You can find various ANSI and ISO standards of Clearance fits in engineering textbooks or machine tool catalogs.

Dimension the GUIDE hole and ROD shaft for a Sliding Clearance fit.  All dimensions below are in millimeters.  Use the following values:

| | | | |
|---|---|---|---|
| Hole | Maximum | 25.021 | |
| | Minimum | 25.000 | |
| Shaft | Maximum | 24.993 | |
| | Minimum | 24.980 | |
| Fit | Maximum | 25.021 - 24.980 = .041 | Max. Hole – Min. Shaft |
| | Minimum | 25.000 – 24.994 = .007 | Min. Hole – Max. Shaft |

Add the maximum and minimum
hole dimensions.  Right-click
∅**25 THRU**.  Click **Properties**
from the Pop-up menu.  Click the
**Tolerance** Tolerance... button
from the Detailing Options dialog
box.

The Dimension Tolerance dialog
box is displayed, Figure 3.50.

Select **Limit** from the Tolerance
Display drop down list.

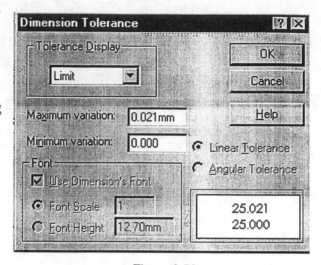

Figure 3.50

Calculate the maximum
variation:

Hole Max: 25.021 – Hole Min: 25.000 = .021 Hole Max.
variation.

Enter **0.021** for the maximum variation.  Enter **0.000** for
minimum variation.  The Max. and Min. dimension values
are displayed in the lower right hand corner.  Display the
tolerance.  Click **OK**, Figure 3.51.

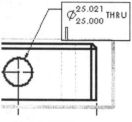

Figure 3.51

Improve view clarity.  Create a gap between the profile
lines and the leader lines.  Create a gap between the center
mark and the leader lines.

Create a gap.  Click the **70** text.  Drag the **leader lines**
outward from the center mark of the circle, Figure 3.52a.
Repeat the above procedure for the **45**, **15**, and **25** linear
dimension text, Figure 3.52b.

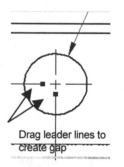

Figure 3.52a

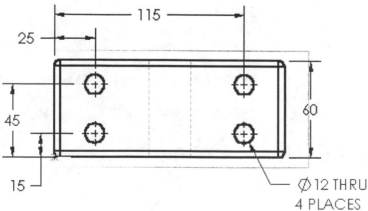

Figure 3.52b

## 3.9    General Notes

For general drawing notes, plan ahead.  Notes provide relative part or assembly information.  Example: Material type, material finish, special manufacturing procedure or considerations, preferred supplier, etc.

Below are a few guidelines to create general notes:

- Use capitol letters

- Use left text justification

- Font size should be the same as the dimension text

Create drawing notes.  Click a **start point** to create the note.  Create the note in the lower left hand corner of the drawing, Figure 3.53.

Place Notes away from the Revision Block

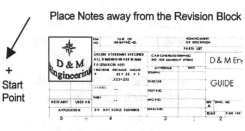

Start
Point

Figure 3.53

Right-click in the **Graphics window**.  Click **Annotations**.  Click **Note**, Figure 3.54a.

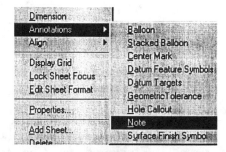

Figure 3.54a

The Note Properties dialog box is displayed, Figure 3.54b.  Type two lines of notes in the Note text box:

Line 1:    **ALL ROUNDS 4 MM**

Line 2:    **REMOVE ALL BURRS**

Display the Note.  Click **OK,** Figure 3.55a.

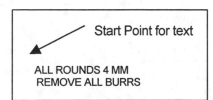

ALL ROUNDS 4 MM
REMOVE ALL BURRS

Figure 3.55a

Use a click and drag procedure to reposition notes on a drawing.

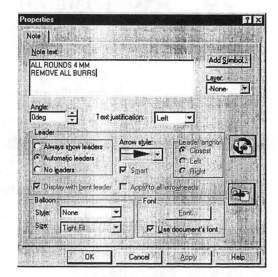

Figure 3.54b

The general note specifies that the radius of all Rounds is 4 millimeters.

Do not double dimension with a note and a dimension. Delete the dimension leader line and text.

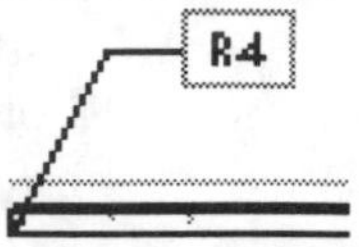

Figure 3.55b

Click the **R4** dimension in *Drawing View1*, Figure 3.55b.

Remove the dimension. Click **Hide**. Refresh the screen. Click the **Repaint** icon.

Create parametric notes by selecting dimensions in the drawing. Example: Specify the maximum width of the GUIDE as a note in the drawing. If the width is modified, the corresponding note is also modified.

Figure 3.56a

Create the note. Right-click and click **Annotations**. Click **Note**. Enter the text, **MAXIMUM WIDTH OF GUIDE** in the Note text box. Click the dimension text **140**, Figure 3.56a. The symbolic representation of the dimension text, "D1@Sketch1@GUIDE-1@DrawingView1" is added to the Note text box. Enter **MM**.

Display the text. Click **OK**, Figure 3.56b.

Align the left edge of the text. Hold the **Ctrl** key down. Click the note, **ALL ROUNDS 4 MM**. Click the note, **MAXIMUM WIDTH OF GUIDE**.

Right-click and click **Align**. Click **Leftmost**, Figure 3.56c.

The notes are leftmost aligned, Figure 3.56d.

ALL ROUNDS 4 MM
REMOVE ALL BURRS

MAXIMUM WIDTH OF GUIDE
140

Figure 3.56b

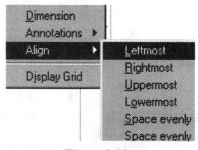

Figure 3.56c

ALL ROUNDS 4 MM
REMOVE ALL BURRS

MAXIMUM WIDTH OF GUIDE
140 MM

Figure 3.56d

Example: Modify the horizontal width dimension to 160 and Rebuild. The parametric Note changes from 140 to 160 to reflect the new value, Figure 3.56e.

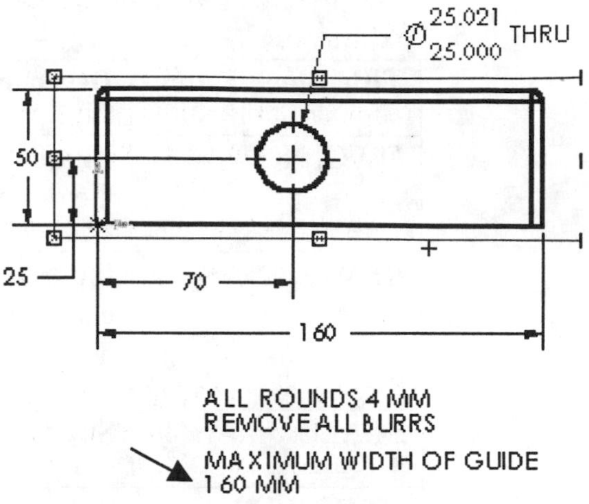

### 3.9.1   Notes in the Title Block

Additional notes are required in the title block. **Zoom in** on the lower right hand section of the title block.

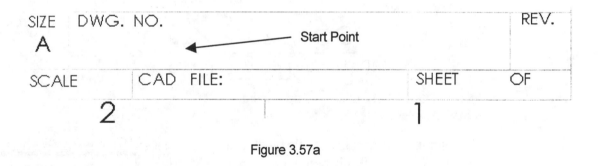

Note: The text box headings: SIZE A, DWG. NO., REV., SCALE, CAD FILE and SHEET OF are entered in the SolidWorks default Sheet Format.

Figure 3.56e

Create a note in the DWG. NO. text box. Click a **start point** in the upper left hand corner below the DWG. NO. text, Figure 3.57a. Right-click in the **Graphics window** and click **Annotation**. Click the **Note**. Enter the drawing number. Enter **56-22222**. Click **OK**.

Figure 3.57a

Enter each note listed in the Title Block Entry table on the next page.

Modify the font size to address smaller text boxes. Remove the **check mark** from the Use document's font. Click the **Font** button, Figure 3.57b. Enter **2.0** mm for the smaller text size.

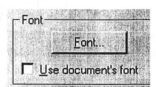

Figure 3.57b

| Title Block Entry | Description | Notes |
|---|---|---|
| DWG. NO. | Unique number to identify and track the drawing | 56 - 22222 |
| REV. | Indicates the number of official changes that were made to the original drawing | A |
| SCALE | Size to which the drawing is created | 1:2 |
| CAD FILE | Software file name | GUIDE.slddrw |
| SHEET | Number of sheets that make up the drawing | 1 OF 1 |

Complete the notes in the title block, Figure 3.58.

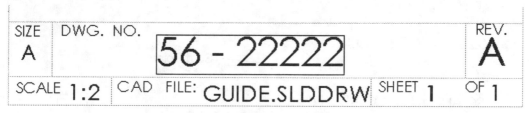

Figure 3.58

The tolerance section in the title block displays information for an English drawing. Create a Metric drawing. Change the tolerance note text.

Edit the sheet format text, "UNLESS OTHERWISE SPECIFIED DIMENSIONS ARE IN INCHES TOLERANCE ARE", Figure 3.59a.

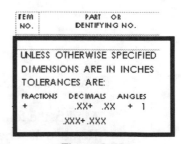

Figure 3.59a

**Zoom in** on the tolerance section of the title block.
Right-click in a **blank area** of the drawing. The mouse pointer displays the Sheet

icon. Click **Edit Sheet format** from the Pop-up menu.

Click on the text: **FRACTIONS DECIMALS ANGLES**. Press the **Delete** key.

Click on the text: **TOLERANCES ARE**. Move the text downward.

Double-click on the text: **DIMENSIONS ARE IN INCHES**. Delete the word **INCHES**. Enter **MILLIMETERS**, Figure 3.59b. Do not abbreviate units in the title block.

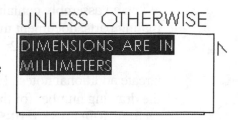

Figure 3.59b

Right-click on the **1 + .XXX. +.XXX + 1** text. The Properties dialog box appears, Figure 3.60.

Enter the following lines:

*   **DECIMALS  ±0.5**

*   **ANGLES  ±0.5**

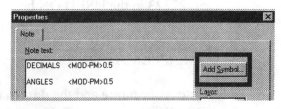

Figure 3.60

Enter the ± symbol from the Modify Symbols list. Click the **Add Symbol** button from the Properties toolbar. Click the **plus/minus** option from the Modifying Symbols drop-down list, Figure 3.61a.

Accept the changes. Click **OK**, Figure 3.61b.

Delete the **XX.XXX** text.

In Project 1, the design team decided that the ROD and GUIDE would be fabricated out of 303 Stainless steel.

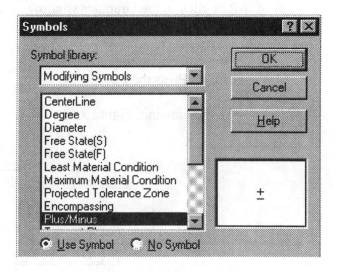

Figure 3.61a

There are numerous types of Stainless steels for various applications. Select the correct material for the application. This is critical!

Double-click the **dash marks** in the MATERIAL text box. Enter **STAINLESS STEEL 303**. Drag the **text** to the center of the MATERIAL text box.

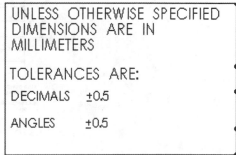

Figure 3.61b

When an assembly contains mating parts, document their relationship, Figure 3.62.

Create additional notes. Enter the drawing number for the GUIDE-ROD ASSY **10-50123** in the NEXT ASSY box.

Enter the ROD drawing number, **56-22223** in the USED ON box.

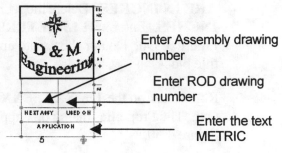

Enter Assembly drawing number

Enter ROD drawing number

Enter the text METRIC

Figure 3.62

Enter the **METRIC** note below in the text APPLICATION.

Enter your **name** and the **date** in the DRAWN box, Figure 3.63.

Right-click in the Graphics window. Click **Edit Sheet**.

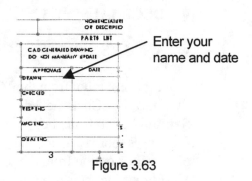

Enter your name and date

Figure 3.63

Fit the drawing to the Graphics window. Press the **f** key.

**Save** the drawing, Figure 3.64.

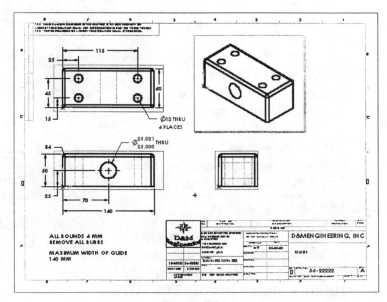

Figure 3.64

### 3.10 Create a Drawing with an Exploded View and a Bill of Materials

Add an *Exploded* view and Bill of Materials to the drawing. Add the GUIDE-ROD assembly *Exploded* view. The Bill of Materials reflects the components of the GUIDE-ROD assembly.

Create a drawing with a Bill of Materials. Perform the following steps:

1. Create a new drawing from the sheet format

2. Display the *Exploded* view of the assembly

3. Insert the *Exploded* view of the assembly into the drawing

4. Label each component

5. Create a Bill of Materials

Let's start.

**Close** all parts and drawings.

Create a new drawing. Use the Custom A-size sheet format. Click the **New** icon from the Standard toolbar. Click **Drawing**. Click **Custom Sheet Format** from the Sheet format dialog box. Select sheet format, **CUSTOM-A**.

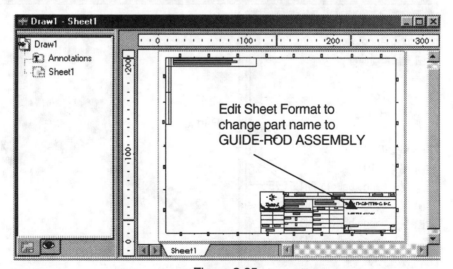

Figure 3.65

Edit the drawing sheet format. Change the part name in the title block. Right-click in the **center** of the drawing. Click **Edit Sheet format** from the Pop-up menu. Double-click on the **part-name text**. Enter **GUIDE-ROD ASSEMBLY** in the Note text box.

Display the new name. Click **OK**, Figure 3.65.

Edit the drawing sheet. Right-click the **center** of the drawing. Click **Edit Sheet** from the Pop-up menu.

Note: Insure that the drawing units are in Millimeters.

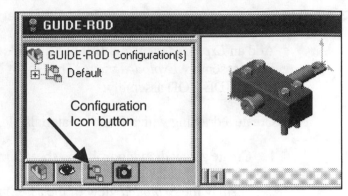

Figure 3.66a

**Save** the drawing. Enter the filename. Enter **GUIDE-ROD**.

Display the *Exploded* view of the assembly. Open the GUIDE-ROD.sldasm assembly. Click the **Open** icon. Click **GUIDE-ROD**.

Explode the view. Click the **Configuration** icon button at the bottom of the ViewManager. Click the **Plus Sign** to the left of the Default entry, Figure 3.66a.

Display the *Exploded* view in the Graphic widow. Double-click the **ExpView1**.

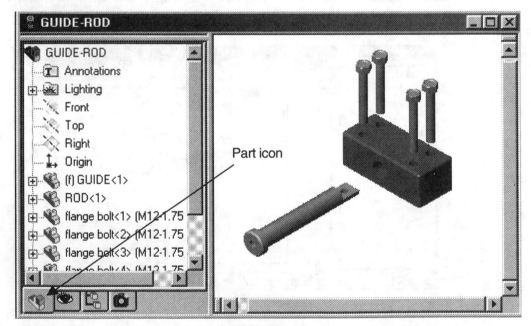

Figure 3.66b

Fit the Exploded view to the Graphics window. Press the **f** key. Click the **Part** icon at the bottom of the View Manager, Figure 3.66b.

Split the Graphics window to view both the assembly and the drawing sheet. Click **Window** from the Main menu. Click **Tile Horizontal,** Figure 3.67.

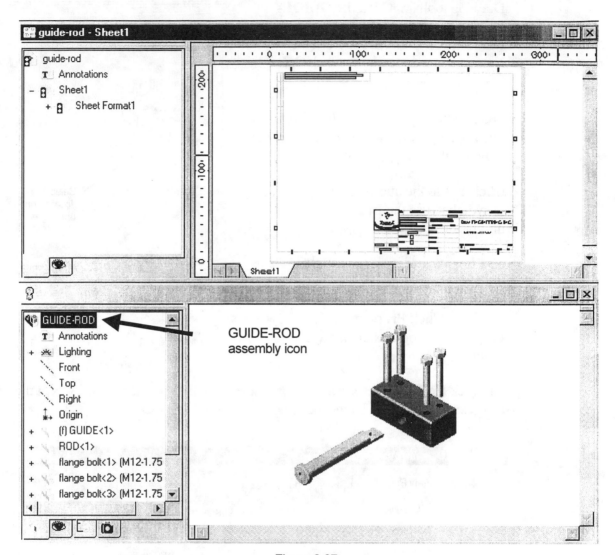

Figure 3.67

Insert the *Exploded* view into the drawing. Click **inside** the drawing sheet. Click the **Named Views** 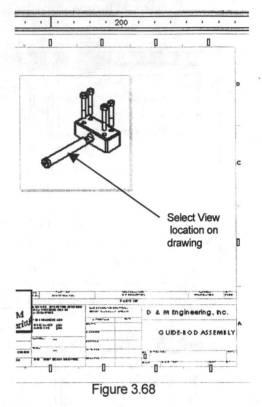 icon from the Drawing toolbar. Click the **GUIDE-ROD assembly** icon in the Assembly window. Click *Isometric* from the Named View dialog box. Click **OK**. The mouse pointer displays a **+** symbol.

Position the view on the right side of the drawing. Click the **inside location** of the drawing sheet, Figure 3.68.

Click **Yes** to the question, "Do you want to switch to use the Isometric true dimensions?"

The *Isometric* view is too small for the drawing sheet. Increase the drawing scale. Right click in the *Isometric* view window. Click **Properties**. The View Properties dialog box is displayed, Figure 3.69.

Figure 3.68

Uncheck the **Use sheet's scale** option. Enter **3** in the Scale text box. Accept the new view scale 1:3. Change the scale of the view. Click **OK**.

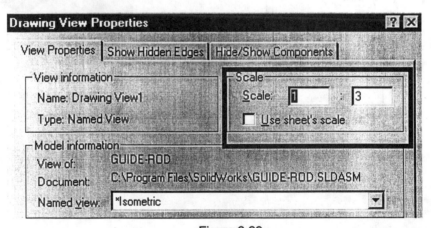

Figure 3.69

**Maximize** the drawing. **Save** the drawing.

Resize the view border. Fit the view to the right half of the drawing.

Drag the **green square handles** of the view boundary, Figure 3.70.

Label each component with a unique item number. The item number is placed inside a circle.

This is called Balloon text.

This Balloon text requires a leader line with a solid arrowhead. Set the arrowhead shape. Click **Tools, Options**. Click the **Document Properties tab**. Click **Arrows**. Set the arrow type for Face/surface. Click the **solid filled arrow** from the Edge/vertex drop down list, Figure 3.71a.

Set the arrowhead size. Enter **1.0** in Height text box. Enter **3.0** for Width. Enter **6.0** for Length.

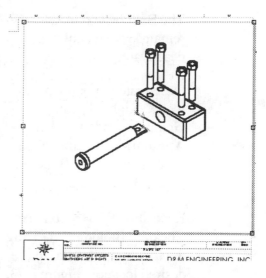

Figure 3.70

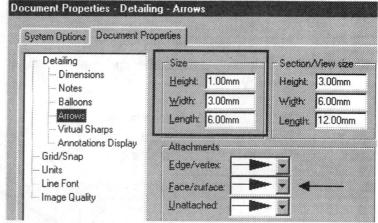

Figure 3.71a

Create the Balloon text. Right-click in the **Graphics window**. Click **Annotations, Balloon**, Figure 3.71b.

Place the first Balloon. Click the **edge** of the GUIDE. A Balloon appears with a leader line and the number 1 inside a circle ⓠ.

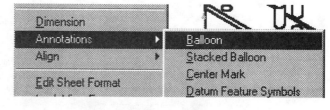

Figure 3.71b

Place the second Balloon. Click the **middle** of the ROD.

Place the third Balloon. Click on the **left front** hex bolt.

The Balloon text mode remains active until it is deactivated.

Deactivate the Balloon text mode. Right-click in the Graphics window. Click **Annotations**. Remove the checkmark. Click **Balloon**, Figure 3.72.

Reposition the Balloon text. Click the **Balloon** text. Drag the **text** by its handles to the desired location, Figure 3.73.

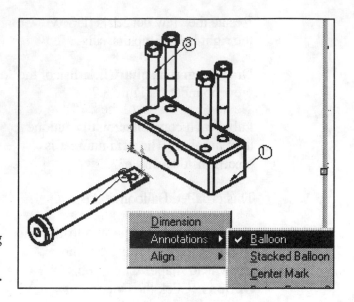

Figure 3.72

Note: Remove Balloon text with the Delete key. Balloons can contain both item number and quantity.

Select the Balloons icon button in the Tools, Options, Document Properties, Balloon dialog box to change format.

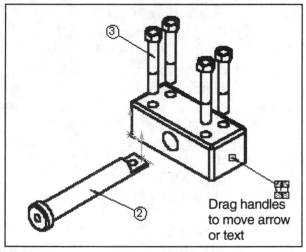

Drag handles to move arrow or text

Figure 3.73

Create a Bill of Materials (BOM). A BOM is created from views containing parts, sub-assemblies or assemblies.

Select the assembly. Click inside the *Isometric* view.

Set the text height of the BOM to equal the dimension text height. Click **Tools** from the Main menu. Click **Options**. Click the **Detailing tab**. Click the **Note Font** icon. Enter text height. Enter **3.5**.

With the drawing view selected, click **Insert** from the Main menu. Click **Bill of Materials**, Figure 3.74. Accept the default Bill of Material template, BOMTEMP.XLS. Click **Open**.

Figure 3.74

The Bill of Materials Properties dialog box is displayed, Figure 3.75a.

Set the text height to 3.0 mm. Click the **Use documents note font when creating the table** checkbox. Uncheck the **Use table anchor point** check box. Click **OK**.

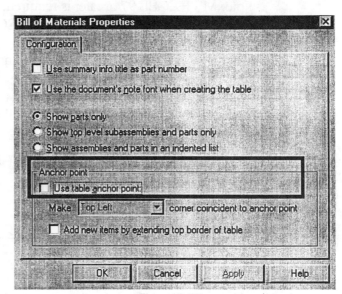

Figure 3.75a

On the Bill of Materials table, the upper left hand corner is called the anchor point. In your custom sheet format, the anchor point for the Bill of Materials overlaps the sheet border. Move the anchor point. Click the **Bill of Materials**. The grip points turn green, Figure 3.75b.

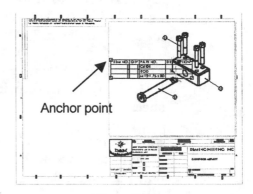

Anchor point

Figure 3.75b

Drag the **Bill of Materials** below the copy write text in the upper left corner, Figure 3.77.

Your custom Bill of Material incorporates:

- ITEM NO.
- QUANTITY
- PART NUMBER
- MATERIAL

| ITEM NO. | QTY. | PART NO. | DESCRIPTION |
|---|---|---|---|
| 1 | 1 | GUIDE | |
| 2 | 1 | ROD | |
| 3 | 4 | M12-1.75 x 80 | |

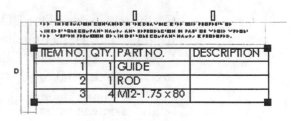

Figure 3.77

The Bill of Materials is created with an EXCEL spreadsheet.

Note: MS EXCEL 97 or later is required. Utilize the default Bill of Material by changing the column heading from DESCRIPTION to MATERIAL.

Run Microsoft EXCEL. Double-click the inside **Bill of Material** table. The EXCEL spreadsheet and toolbars are displayed. EXCEL defines entries in the table by Cells. Each cell is labeled by a row letter and a column number. Example: Text DESCRIPTION is in cell D1.

Edit the text. Click the **DESCRIPTION** text in cell D1. Enter **MATERIAL** in cell D1. Enter **SS 303** in the MATERIAL column for the GUIDE. This is in location D2. Enter **SS 303** in the MATERIAL column for the ROD. This is in location D3. Enter **ALUMINUM** for the metric hex bolts. This is in location D4, Figure 3.78.

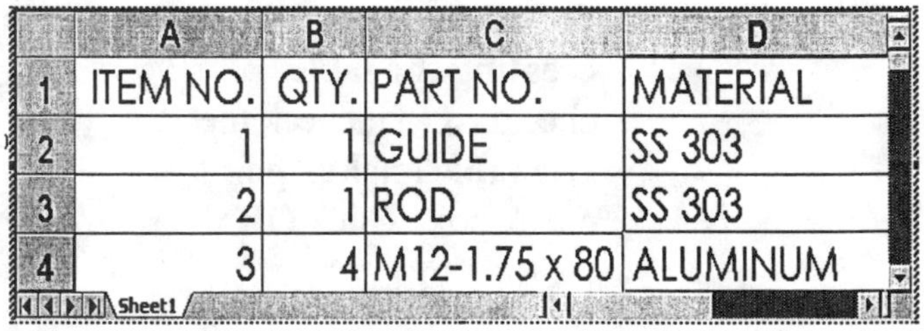

| | A | B | C | D |
|---|---|---|---|---|
| 1 | ITEM NO. | QTY. | PART NO. | MATERIAL |
| 2 | 1 | 1 | GUIDE | SS 303 |
| 3 | 2 | 1 | ROD | SS 303 |
| 4 | 3 | 4 | M12-1.75 x 80 | ALUMINUM |



Figure 3.78

Return to SolidWorks. Exit EXCEL. Click the **drawing sheet**.

**Save** the drawing. The Bill of Materials is saved with the drawing, Figure 3.79.

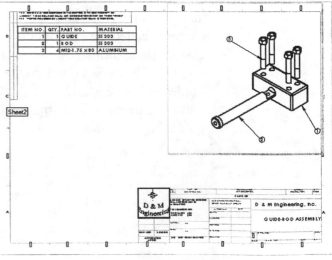

Figure 3.79

### 3.11 Associative Part, Assembly and Drawing

The associative part, assembly and drawing share a common database. Verify the association between the part, assembly and drawing.

Modify a dimension on the drawing. Open the GUIDE drawing. Click the **Open** 📂 icon. Enter **GUIDE**.

Double-click on the vertical linear dimension **50** in *Drawing View1* (Front). Enter **80** in the Modify dialog box. Click **Rebuild**. All views are updated, Figure 3.80.

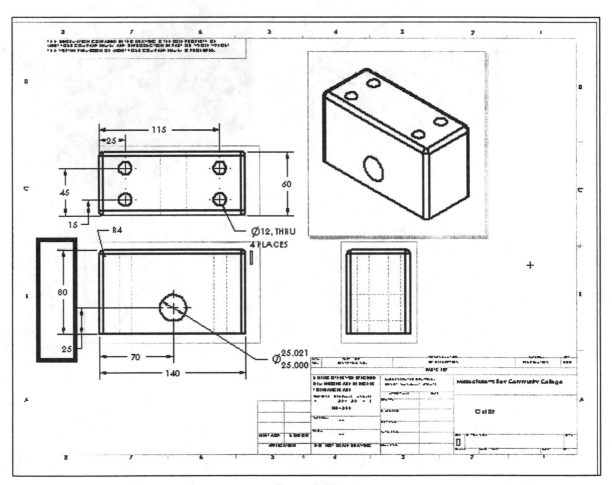

Figure 3.80

View the updated part. **Open** the GUIDE. The system provides a prompt. The file is changed and it requires an update. Click **OK**, Figure 3.81. View the updated assembly. **Open** the GUIDE-ROD assembly, Figure 3.82.

Drawings are an integral part of the design process. Part, assemblies and drawings all work together. From your initial design concepts, you created parts and drawings that fulfill the design requirements of your customer.

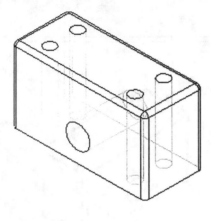

Figure 3.81

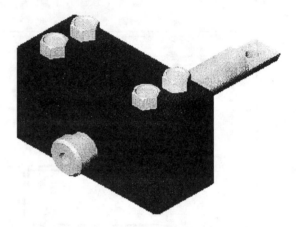

Figure 3.82

### 3.12 Questions

1.  What is a Bill of Materials?

2.  Name the two major design modes used to develop a drawing in SolidWorks?

3.  What are the three various sheet format options in SolidWorks?

4.  Name seven components that are commonly found in a title block.

5.  How do you add an Isometric view to the drawing?

6.  In SolidWorks, drawing file names end with a _____ suffix?

7.  True or False.  In SolidWorks, if a part is modified, the drawing is automatically updated.

8.  True or False.  In SolidWorks, when a dimension in the drawing is modified, the part is automatically updated.

9.  Name three guidelines to create General Notes.

10. True or False.  Most engineering drawings us the following font: Time New Roman – All small letters.

11. What are Leader lines?

12. Name the three ways that Holes and other circular geometry can be dimensioned.

13. What are Center Marks?

14. How do you calculate the maximum and minimum variation?

## 3.13 Exercises

Exercise 3.1     Create a drawing for the U-CHANNEL.  The part, U-CHANNEL was created in Project 1, Exercise 1.4.

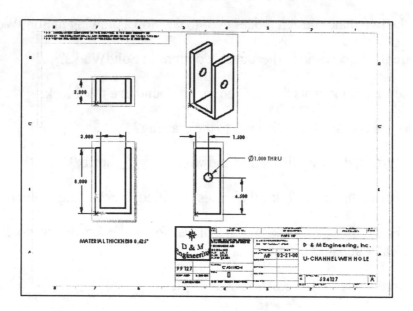

Exercise 3.2     Create an assembly drawing and Bill of Materials for Project 2, Exercise 2.3.

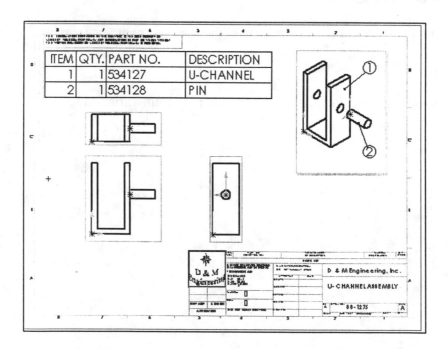

Exercise 3.3     Create a Metric drawing from the part, PLATE. Set Metric units in the Grid/Units options. Insert only the *Top* view into the drawing. Rotate the *Top* view. Modify the dimensions. Add Notes, Hole Call Out and Center Marks.

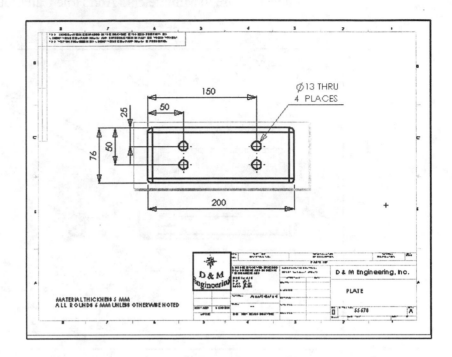

Exercise 3.4     Create an assembly drawing and Bill of Materials for the two PLATEs and four STANDOFFS.

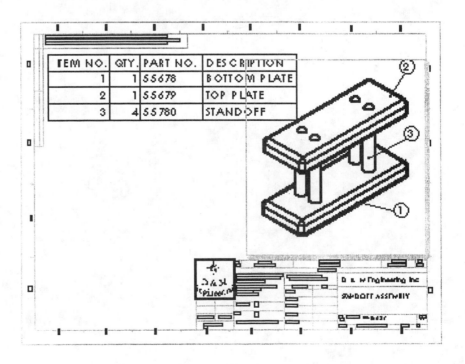

Exercise 3.5    Create a drawing for the ROD.

Rotate the views for the correct orientation.

Dimensions are in Millimeters.   Add Hole Call Outs and Notes.

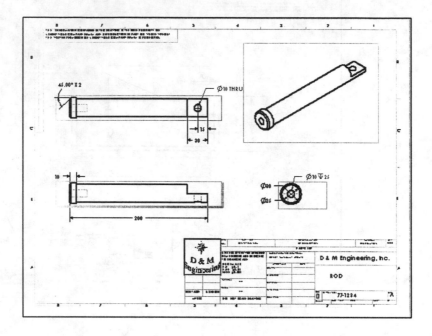

# Project 4

## Extrude and Revolve Features

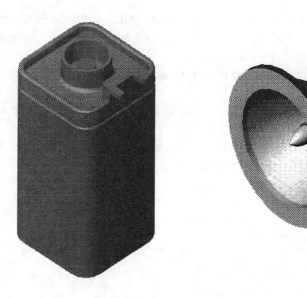

Below are the desired outcomes and usage competencies based upon the
completion of Project 4.

| Project Desired Outcomes | Usage Competencies |
|---|---|
| • A comprehensive understanding of the customer's design requirements and desires. | • To comprehend the fundamental definitions and process of Feature-Based 3D Solid Modeling. |
| • A product design that is cost effective, serviceable and flexible for future manufacturing revisions. | • Specific knowledge and understanding of the Extrude and Revolve features. |
| • Four key flashlight components. | |

## 4    Project 4 – Extrude and Revolve Features

### 4.1    Project Objective

Create four components of the flashlight.  Create the BATTERY, BATTERY PLATE, LENS and BULB.

### 4.2    Project Situation

You are employed by a company that specializes in providing promotional trade show products.  Your company is expecting a sales order for 100,000 flashlights with a potential for 500,000 units next year.  Prototype drawings of the flashlight are required in three weeks.

You are the design engineer responsible for the project.  You contact the customer to discuss design options and product specifications.  The customer informs you that the flashlights will be used in an international marketing promotional campaign.  Key customer requirements:

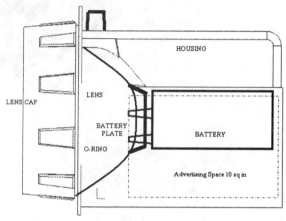

- Inexpensive reliable flashlight

- Available advertising space of 10 square inches

- Light weight semi indestructible body

- Self standing with a handle

Your company's standard product line does not address

Figure 4.1

the above key customer requirements.  The customer made it clear that there is no room for negotiation on the key product requirements.

You contact the salesperson and obtain additional information on the customer and product.  This is a very valuable customer with a long history of last minute product changes.  The job has high visibility with great future potential.

In a design review meeting, you present a conceptional sketch.  Your colleagues review the sketch.  The team's consensus is to proceed with the conceptual design, Figure 4.1.

The first key design decision is the battery. The battery type will directly affect the flashlight body size, bulb intensity, case structure integrity, weight, manufacturing complexity and cost.

You review two potential battery options:

- A single 6-volt lantern battery

- Four 1.5 volt D cell batteries

The two options effect the product design and specification. Think about it.

A single 6-volt lantern battery is approximately 25% higher in cost and 35% more in weight. The 6-volt lantern battery does provide higher current capabilities and longer battery life.

A special battery holder is required to incorporate the four 1.5 volt D cell configuration. This would directly add to the cost and design time of the flashlight, Figure 4.2.

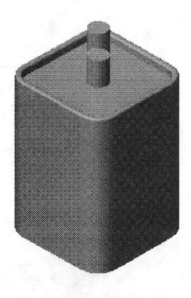
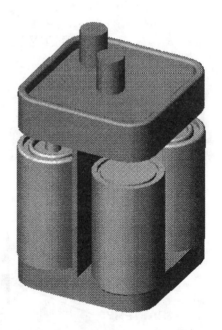

Figure 4.2

Time is critical. For the prototype, you decide to use a standard 6-volt lantern battery. This eliminates the requirement to design and procure a special battery holder. However, you envision the 4-D cell battery model for the next product revision. You design the flashlight to accommodate both battery design options.

Battery dimensional information is required for the design. Where do you go? Potential sources: product catalogs, company web sites, professional standards organizations, design handbooks and colleagues.

The team decides to purchase the following components: 6-volt BATTERY, LENS ASSEMBLY, SWITCH and an O-RING. Your company will design and manufacture the following components: BATTERY PLATE, LENSCAP, HOUSING and SWITCH PLATE.

| Purchased Parts | Designed Parts |
|---|---|
| BATTERY | BATTERY PLATE |
| LENS ASSEMBLY | LENS CAP |
| SWITCH | HOUSING |
| O-RING | SWITCH PLATE |

## 4.3    Project Overview

Create four parts in this section, Figure 4.3a:

- BATTERY

- BATTERY PLATE

- LENS

- BULB

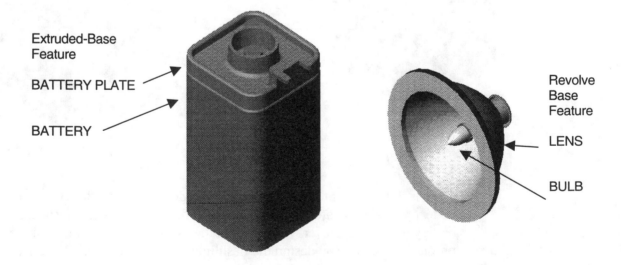

Figure 4.3a

Two major *Base* features are discussed in this project:

- *Extrude* – BATTERY and BATTERY PLATE

- *Revolve* – LENS and BULB

Note: Dimensions and features are used to illustrate the SolidWorks functionality in a design situation. Wall thickness and thread size have been increased for improved picture illustration. Parts have been simplified.

You will create four additional parts in Project 5 for a final flashlight assembly, Figure 4.3b.

- O-RING

- LENSCAP

- SWITCH

- HOUSING

Figure 4.3b

## 4.4  BATTERY

The BATTERY is a simplified representation of an OEM component. The BATTERY consists of the following features:

- *Extruded-Base*

- *Extruded-Cut*

- *Edge Fillets*

- *Face Fillets*

The battery terminals are represented as cylindrical extrusions. The battery dimension is obtained from the ANSI standard 908D.

Note: A 6-volt lantern battery weighs approximately 1.375 pounds (.62kg). Locate the center of gravity closest to the center of the battery.

### 4.4.1    BATTERY Overview

Create the BATTERY, Figure 4.4a.  Identify the required BATTERY features.

- *Extruded-Base*.  The *Extruded-Base* feature is created from a symmetrical square sketch, Figure 4.4b.

- *Fillet*.  The *Fillet* feature is created by selecting the vertical edges and the top face, Figure 4.4c, Figure 4.4e.

- *Extruded-Cut*.  The *Extruded-Cut* feature is created from the top face offset, Figure 4.4d.

- *Boss-Extrude*.  The *Boss-Extrude* feature is created to represent the battery terminals, Figure 4.4f.

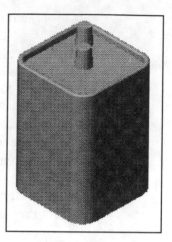

Figure 4.4a

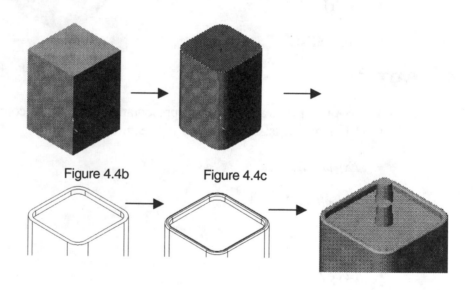

Figure 4.4b            Figure 4.4c

Figure 4.4d            Figure 4.4e            Figure 4.4f

Let's create the BATTERY.

### 4.4.2    Create the English Template

Create an English document template.  Click the **New** icon.  Click the **Part** template.  Click **OK**.

Rename the reference planes *Front*, *Top* and *Right*, Figure 4.5.

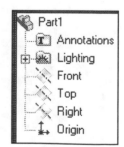

Figure 4.5

Set the System Options. Click **Tools**, **Options**, from the Main menu. The System Options - General dialog box is displayed. Click the **Input dimension value** check box from the General box, Figure 4.6a.

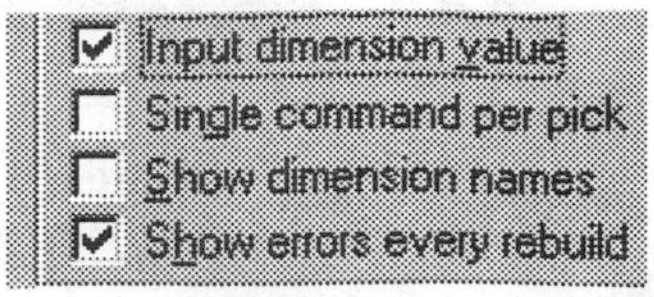

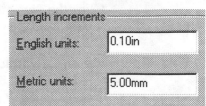

Figure 4.6a

Set the English units Length increment. Click the **Spin Box Increments** option. Click the English units **text box**. Enter **0.10**, Figure 4.6b.

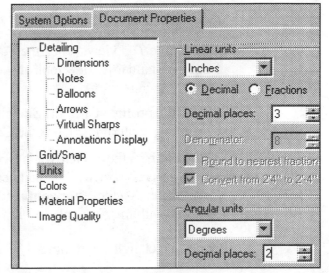

Figure 4.6b

Set the Document Properties. Click the **Document Properties** tab. Click the **Units** option. Enter **Inches** from the Linear units list box. Click the **Decimal** button. Enter **3** in the Decimal places spin box, Figure 4.7a.

Figure 4.7a

Set the sketch Grid/Snap. Enter **1.000** in the Major grid spacing spin box. Enter **5** in the Minor-lines per major text box, Uncheck all the **Snap check boxes**, Figure 4.7b.

Accept the new settings. Click **OK** from the Document Properties dialog box.

Click **File** from the Main menu. Click **Save As**. Click **\*.prtdot** from the Save As type list box. The default Templates file folder is displayed. Enter **PARTENGLISH TEMPLATE** in the File name text box. Click **Save**.

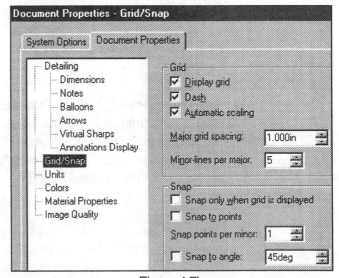

Figure 4.7b

### 4.4.3   Create the BATTERY

Create the BATTERY.  Click the **New** ⬜ icon.  Click
**PARTENGLISHTEMPLATE** from the Template dialog box.  Click **OK**.

Click the **Save** 💾 icon.  Enter the part name.  Enter **BATTERY**.  Click
**SAVE**.

### 4.4.4   Create an Extruded-Base Feature

The *Extruded-Base* feature uses a square sketch drawn symmetrical on the
*Top* reference plane.  Symmetrical relationships require centerlines.

Select the *Sketch* plane.  Click the *Top* plane from the FeatureManager.

Display the *Top* view.  Click the **Top** 🔲 icon
from the Standards View toolbar.

Sketch the profile.  Click the **Sketch** ✏ icon.
Create two centerlines.  Click the **Centerline**
▮ icon from the Sketch Tools toolbar.
Sketch a **vertical centerline** collinear with the
*Right* plane.  Sketch a **horizontal centerline**
collinear with the *Front* plane, Figure 4.8.

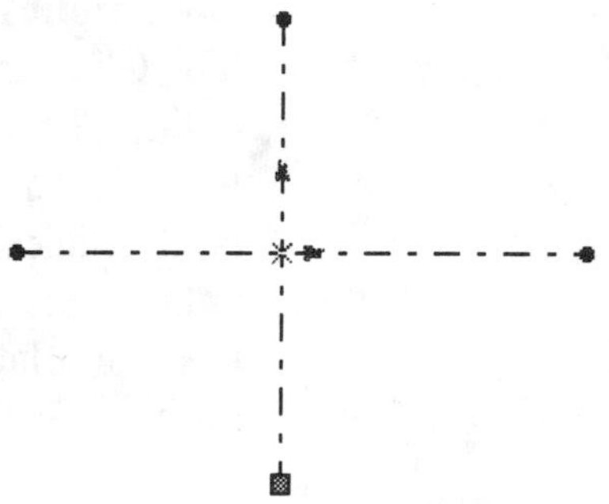

Figure 4.8

Create a rectangle approximately centered

about the *Origin* ⌞.  Click the **Rectangle**
▦ icon.  Click the **first point** in the lower
left quadrant.  Drag the **mouse pointer** to the
upper right quadrant.  Create the second
point.  Release the **mouse button**,
Figure 4.9.

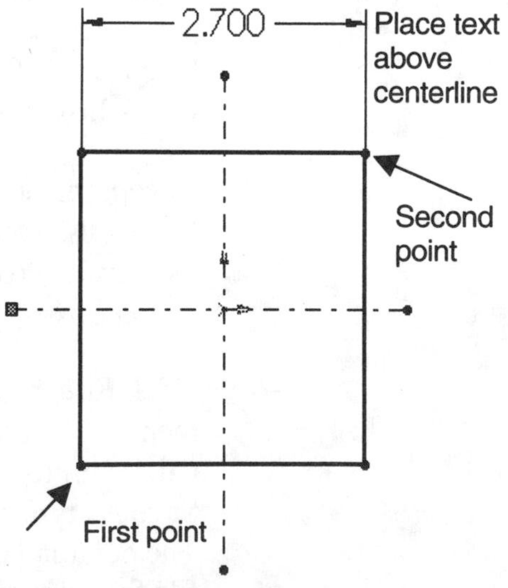

Add dimensions.  Click the **Dimension** ✏
icon.  Click the **top horizontal line**.  Drag
the **mouse pointer** off the *Sketch*.  Position
the dimension text.  Click the **text location**
above the vertical centerline.  Enter **2.700** for
width.

Note: The part precision is set to 3 decimal
places.  Example: 2.700 is displayed.  If you
enter 2.7, the value 2.700 is displayed.  For consistency, the dimension values
for the text include the number of decimal places required.  The units are set to
inches.

2.700 — Place text above centerline

Second point

First point

Figure 4.9

Add geometric relationships.  Click the **Add Relations** ⊥ icon.

Note: The Line# may be different than the numbers above.  The Line# is dependent on the creation order.

Add a symmetric relation.  Click the vertical centerline, **Line1**.  Click the two vertical profile lines, **Line4** and **Line6**.  Click the **Symmetric** button.  Click **Apply**, Figure 4.10.

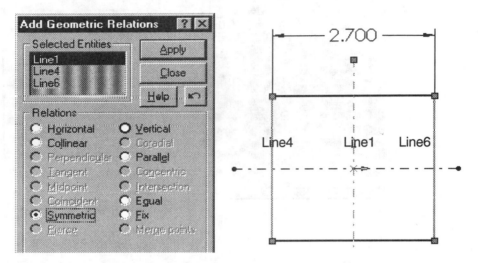

Figure 4.10

Add a symmetric relation.  Click the horizontal centerline, **Line2**.  Click the two horizontal profile lines, **Line3** and **Line5**.  Click the **Symmetric** button. Click **Apply**, Figure 4.11.

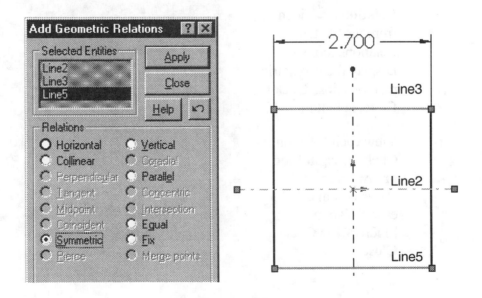

Figure 4.11

Add an equal relation.  Click the top horizontal profile line, **Line3**.  Click the left vertical profile line, **Line4**.  Click the **Equal** button.  Click **Apply**.  Click **Close**, Figure 4.12.  The black *Sketch* is fully defined.

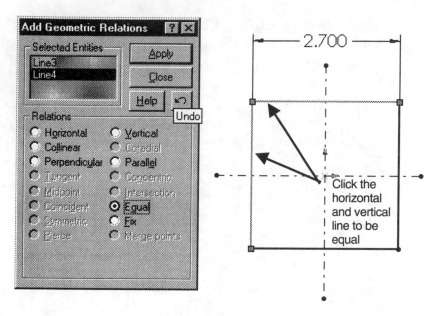

Figure 4.12

Click the **Display/Delete Relations** 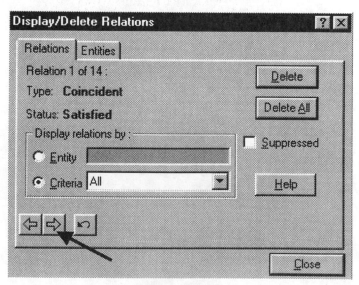 icon from the Sketch Relations toolbar. Display the geometric relationships.  Click the **Criteria** button.

View each Relation. Click the **right blue arrow**.  Repeat the above commands for each Relation, Figure 4.13.  Click **Close**.

Figure 4.13

Extrude the *Sketch*. Click the
**Extruded Boss/Base** 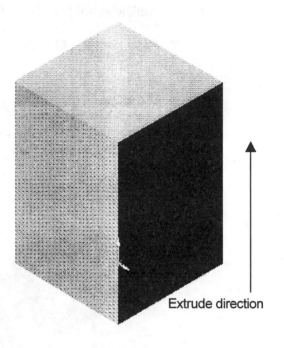 icon.
Blind is the default Type option.
Enter **4.100** for Depth. Display
the *Base-Extrude* feature. Click
**OK**, Figure 4.14.

Click the **Zoom to Fit** icon.

**Save** the BATTERY.

### 4.4.5    Create the Fillets

The vertical sides on the
BATTERY are rounded. Use the
*Fillet* feature to round the 4 sides.

Display Hidden Lines Gray.
Click the **Hidden In Gray**
icon.

Figure 4.14

Create a *Fillet* feature. Click the **Fillet** icon from the Feature toolbar.
Click the **4 vertical edges**, Figure 4.15. Enter **0.500** for Radius. Display the
*Fillet* feature. Click **OK**.

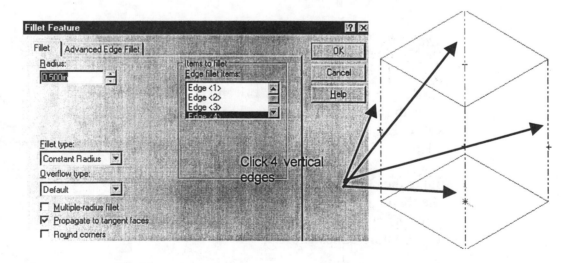

Figure 4.15

Rename *Fillet1* to *Side-Fillets* in the Feature Manager.

**Save** the BATTERY.

### 4.4.6    Extruded-Cut Feature with Offset Face

The *Cut* feature removes material. An Offset Edge takes existing geometry, extracts it from an edge or face and locates it on the current sketch plane. Create a *Cut* feature. Offset the existing *Top* face.

Select the *Sketch* plane. Click the *Top* face, Figure 4.16. Display the face. Click the **Top** icon from the Standards View toolbar, Figure 4.17.

Figure 4.16

Figure 4.17

Create the *Sketch*. Click the **Sketch** icon. Offset the existing geometry. Click the **Offset** icon from the Sketch Tools toolbar. The Offset Entities dialog box is displayed, Figure 4.18.

Enter the Offset. Enter **0.150**. Click the **Reverse** check box. Display the new Offset inside the original profile. Click **Apply**. Click **Close**.

Verify that the Offset is inside the original profile.

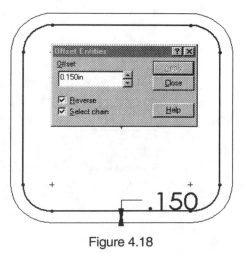

Figure 4.18

Extrude the Offset profile. Click the **Extrude-Cut** 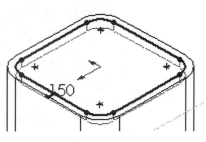 icon from the Feature toolbar. The Extrude Cut Feature dialog box is displayed. Display the profile. Click the **Isometric** icon from the Standards View toolbar, Figure 4.19.

Enter **0.200** for Depth. Display the *Extruded-Cut*. Click **OK**, Figure 4.20.

Figure 4.19

Rename *Cut-Extrude1* to *Top-Cut*.

**Save** the BATTERY.

**4.4.7    Face Fillet**

Both top outside edges require fillets. Use a face to create a *Fillet* feature.

Figure 4.20

Display the Selection Filter toolbar. Click **View** from the Main menu. Click **Tools**, **Selection Filter**.

Select the top thin face. Click the **Face Filter** icon from the Selection Filter toolbar. Click the **top thin face**. Click the **Fillet** icon. Face<1> is displayed in the Edge fillet items box. Enter the *Fillet* Radius. Enter **0.050**, Figure 4.21.

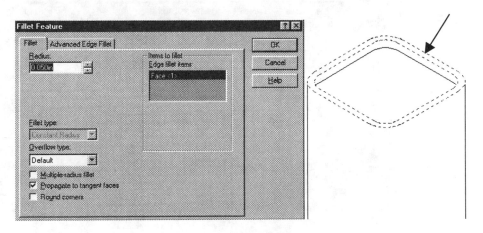

Fillet 4.21

Display the *Fillet* on both top edges. Click
**OK**, Figure 4.22.

Turn off the Face Filter. Click the **Face
Filter** icon.

Rename *Fillet2* to ***TopEdge-Fillet***.

**Save** the BATTERY.

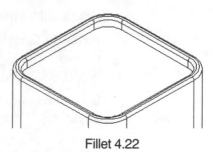

Fillet 4.22

Note: Do not select a *Fillet* radius which is larger that the surrounding
geometry.

Example: The top edge face width is 0.150". The *Fillet* is created on both
sides of the face. A common error is to enter a *Fillet* too large for the existing
geometry. A minimum face width of 0.200" is required for a *Fillet* radius of
0.100".

The following error occurs went the *Fillet* radius is too large for the existing
geometry, Figure 4.23.

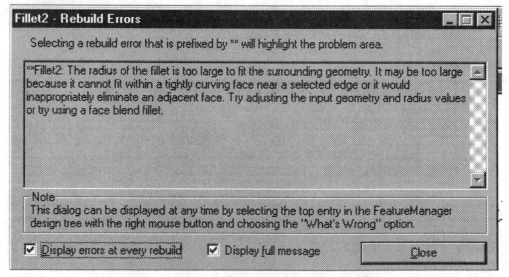

Figure 4.23

Avoid the *Fillet* Rebuild error. To avoid this error, reduce the *Fillet* size or
increase the face width.

**4.4.8        Battery Terminals**

Two terminals are required.  To conserve design time, represent the terminals as an extruded cylinder.

Obtain the specific terminal location from the battery standard.

Select the *Sketch* plane.  Click the **face** of the *Top-Cut* feature, Figure 4.24.

Figure 4.24

Display the *Sketch* plane.  Click the **Top** icon from the Standards View toolbar.

Sketch the Profile.  Click the **Sketch** icon. Sketch a circle.  Click the **Circle** icon.  Click the **center point** of the circle coincident to the

*Origin* .  Click the **second point**.

Click the **Dimension** icon.  Click the **circumference** of the circle.  Click the **text location**.  Enter **0.500**, Figure 4.25.

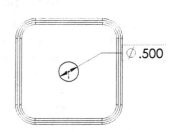

Figure 4.25

Copy the sketched circle.  Click the **Select** icon. Hold the **Ctrl** key down.  Click the **diameter** of the circle.  Drag the **circle** to the upper left quadrant. Create the second circle.  Release the **mouse button**, Figure 4.26a.  Release the **Ctrl** key.

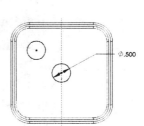

Figure 4.26a

Add Relationships.  Click the **Add Relations** icon.  Add an equal relation.  Click the **circumference of each circle**, Figure 4.26b.  Click **Equal**.  Click **Apply**.  Click **Close**.

By default, the PropertyManager is displayed when the Sketch is active.  The FeatureManager is displayed when the Part is active.

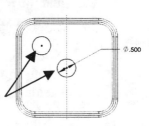

Figure 4.26b

Display the FeatureManager. Click the **Part** 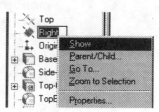 icon in the lower left corner.

The dimension between the center points is critical. Dimension the distance between the two center points with an aligned dimension.

The *Right* plane is the dimension reference. Right-click the ***Right*** plane from the FeatureManager. View the plane. Click **Show**, Figure 4.27.

Figure 4.27

Display the PropertyManager. Click the **PropertyManager** icon.

Add dimensions. Click the **Dimension** icon. Click the **two center points** of the circles. Drag the **dimension text** off the profile. Release the **mouse button**, Figure 4.28a.

Enter **1.000** for the aligned dimension.

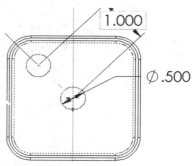

Figure 4.28a

The dimension text toggles between linear and aligned. An aligned dimension is created when the dimension is positioned between the two circles.

Create an angular dimension. Click the **Centerline** icon. Sketch a centerline between the **two circle center points**, Figure 4.28b. Create an acute angular dimension. Click the **Dimension** icon. Click the **centerline** between the two circles. Click the ***Right*** plane.

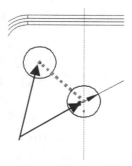

Figure 4.28b

Drag the **dimension text** between the centerline and the *Right* plane off the profile. Release the **mouse button**. Enter **45**, Figure 4.28c.

Note: Acute angles are less than 90 degrees. Acute angles are the preferred dimension standard.

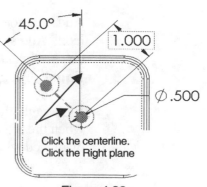

Click the centerline.
Click the Right plane

Figure 4.28c

The overall battery height is a critical dimension. The battery height is 4.500". Calculate the depth of the extrusion:

4.500" – (4.100" *Base-Extrude* height – 0.200" Offset cut depth) = 0.600". The depth of the extrusion is 0.600".

Extrude the *Sketch*. Click the **Extrude Boss/Base** icon. Blind is the default Type option. Enter **0.600** for Depth from the Extrude Feature dialog box, Figure 4.29a.

Create a truncated cone shape for the battery terminals. Check the **Draft While Extruding** box.

A draft angle is a taper. Enter **5** in the Angle text box. Display the *Boss-Extrude* feature. Click **OK**, Figure 4.29b.

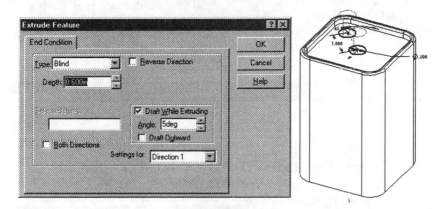

Figure 4.29a

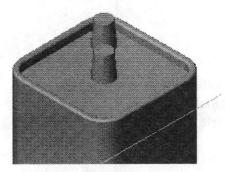

Figure 4.29b

Verify the overall height. Click **Tools**, **Measure** from the Main menu. Click the **Right** ⊞ icon from the Standard Views toolbar. Click the **top edge** of the battery terminal. Click the **bottom edge** of the battery, Figure 4.30. The overall height, Delta Y is 4.500". Click **Close**.

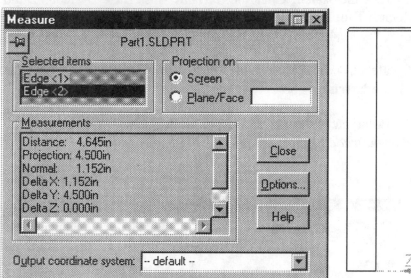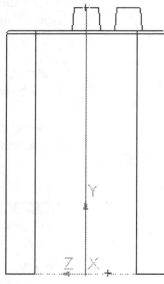

Figure 4.30

Rename *Boss-Extrude1* to *Terminals*, Figure 4.31.

Hide all planes. Click **View** from the Main menu. Click Planes.

Display the *Trimetric* view. Click the **View Orientation** 🖉 icon. Double-click **Trimetric**.

**Save** BATTERY.

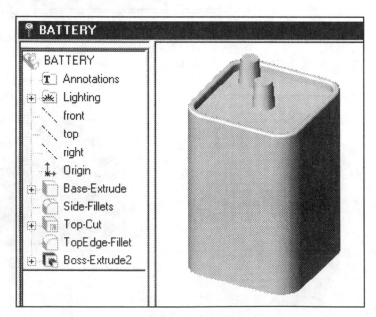

Figure 4.31

**4.5     BATTERY PLATE**

The BATTERY PLATE
has a variety of design
functions.  It:

- Aligns the LENS
  assembly.

- Creates an electrical
  connection between
  the SWITCH
  assembly,
  BATTERY and
  LENS.

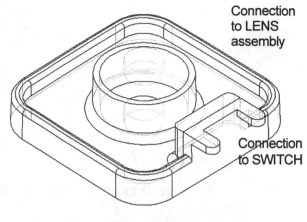

Connection
to LENS
assembly

Connection
to SWITCH

Figure 4.32

Design the
BATTERY PLATE.

Utilize features from the BATTERY to develop the BATTERY PLATE.

**4.5.1     BATTERY PLATE Overview**

Create the BATTERY PLATE.  Modify the
BATTERY features.  Create two holes from
the original sketched circles.  Use the
*Extruded-Cut* feature, Figure 4.33.

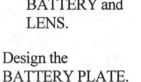

Figure 4.33

Modify the dimensions of the *Base* feature.
Add a 1-degree draft angle.

Note: A sand pail contains a draft angle.
The draft angle assists the sand to leave the pail
when the pail is flipped upside down.

Create a new *Extruded-Thin* feature.  Offset the
center circular sketch, Figure 4.34.

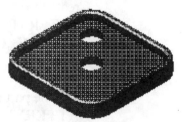

The *Extruded-Thin* feature contains the LENS.
Create an inside draft angle.  The draft angle
assists the LENS into the *Holder*.

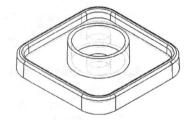

Figure 4.34

Create an *Extruded-Boss* feature using two depth directions, Figure 4.35.
Create an *Extruded-Boss* feature using sketched mirror geometry, Figure 4.36.

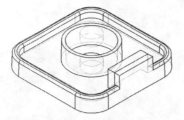

Figure 4.35

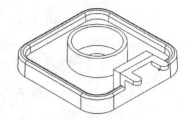

Figure 4.36

Create *Face* and *Edge Fillet* features to remove sharp edges, Figure 4.37.

Let's create the part, BATTERYPLATE.

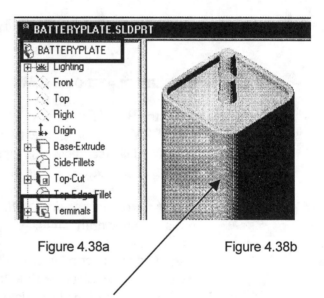

Figure 4.37

### 4.5.2    Create a New Part from an Existing Part

Create the BATTERYPLATE from the BATTERY. Click **File**, **SaveAs**.

Enter the part name. Enter **BATTERYPLATE**. Click **Save**. The BATTERYPLATE part icon is displayed at the top of the FeatureManager, Figure 4.38a.

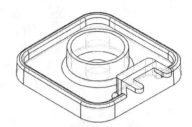

Figure 4.38a                               Figure 4.38b

### 4.5.3    Delete and Modify Features

Create two holes. Delete the *Terminals* feature and reuse the circle sketch. Click the *Terminals* feature, Figure 4.38b.

Remove the *Terminals* feature. Click **Edit** from the Main menu. Click **Delete**. Click **Yes** from the Confirm Delete dialog box. Do not delete the circle sketch, Figure 4.39.

Create an *Extruded-Cut* feature from the two circles.

Click *Sketch3* from the FeatureManager. Click the **Cut-Extrude** 🔲 icon. Click **Through All** for the Depth. Create the cut holes. Click **OK**, Figure 4.40.

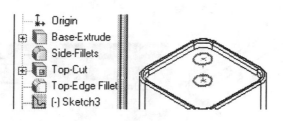

Figure 4.39

Figure 4.40

Rename *Cut-Extrude* to *Holes*. **Save** the BATTERYPLATE.

Modify the *Base-Extrude* feature. Right-click the *Base-Extrude* feature. Click **Edit Definition** from the Pop-up menu. Change the overall Depth to **0.400**, Figure 4.41.

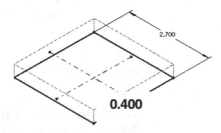

Figure 4.41

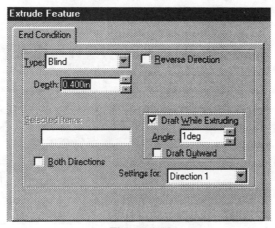

Figure 4.42

Click the **Draft While Extruding** check box. Enter **1** in the Angle text box, Figure 4.42.

Display the modified *Base* feature, Figure 4.43. Click **OK**.

Figure 4.43

**Save** the BATTERYPLATE.

Turn the grid snap off. Click **Tools** from the Main menu. Click **Options**.
Click **Document Properties-Grid**. Uncheck the **Snap to points** check box.
Click **OK**.

### 4.5.4   Create the Extruded-Thin Feature

The *Holder* is created with a circular *Extruded-Thin* feature. Create the
*Holder* feature.

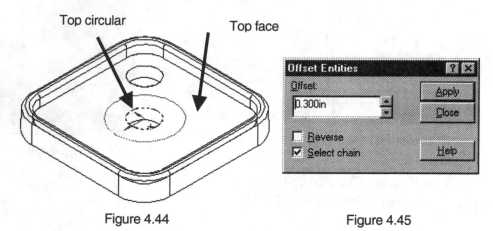

Top circular                                     Top face

Figure 4.44                                      Figure 4.45

Select the *Sketch* plane. Click the **top face**, Figure 4.44.

Sketch the profile. Click the **Sketch** icon. Click the **top circular edge** of
the center *Hole* feature. Click the **Offset Entities** icon. Enter **0.300**,
Figure 4.45. Click **Apply**. Click **Close**.

Extrude the *Sketch*.
Click the **Extruded
Boss/Base** icon.
Click the **Thin**

**Feature** option from
the Extrude as list
box, Figure 4.46.

Blind is the default
Type option. Enter
**0.400** for Depth.
Click the **Draft
While Extruding**
check box. Enter **1**
in the Angle text
box.

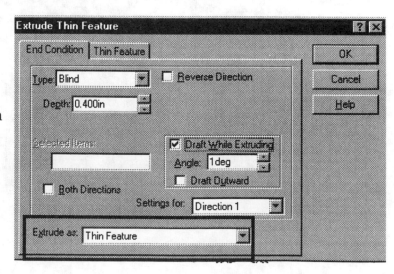

Figure 4.46

Click the **Thin Feature** tab, Figure 4.47. The dialog box name changes to Extrude Thin Feature. The Thin Feature tab appears.

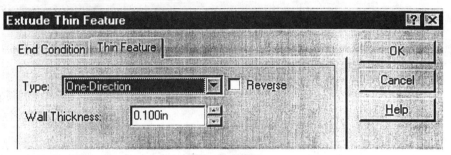

Figure 4.47

Enter **One-Direction** for Type. Enter **0.100** for Wall Thickness. Display the *Thin* feature. Click **OK**, Figure 4.48.

Rename *Boss-Extrude-Thin1* to *Holder*.

**Save** the BATTERYPLATE.

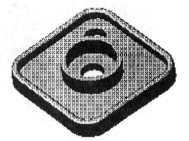

**4.5.5**  **Create a Connector Switch with the Extruded-Boss Feature**

Figure 4.48

The next two *Extruded-Boss* features are used to connect the BATTERY to the SWITCH.

Create the first *Extruded-Boss* feature. Extrude the *Sketch* in two directions.

**Zoom** and **Rotate** the view, Figure 4.49a.

Select the *Sketch* plane. Click the **inside right face**, Figure 4.49b.

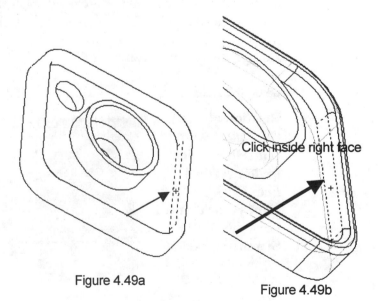

Figure 4.49a

Figure 4.49b

Display the *Right* view.

Click the **Right** icon, Figure 4.50.

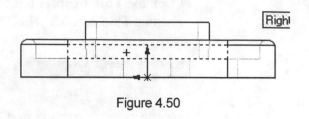

Figure 4.50

Sketch the profile. Click the **Sketch** icon. Click the **Centerline** icon.

Sketch a **vertical centerline**, coincident with the *Origin*, Figure 4.51.

Click the **Rectangle** icon. Sketch a **rectangle** coincident with the bottom and top edges on either side of the centerline.

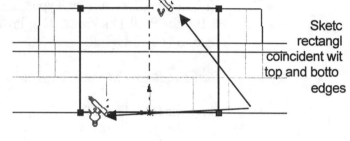

Sketc rectangl coincident wit top and botto edges

Figure 4.51

Geometric relationships are captured as you sketch.

The mouse pointer icon displays the following relationships: Horizontal, vertical, coincident, midpoint, intersection, tangent and perpendicular.

Note: If Automatic Relations are not displayed, Click Tools from the Main menu. Click Options, General, Automatic Relations in the Sketch box.

Add geometric relationships.

Click the **Add Relations** icon. Click the **centerline**. Click the **left vertical line**. Click the **right vertical line**. Click the **Symmetric** button.

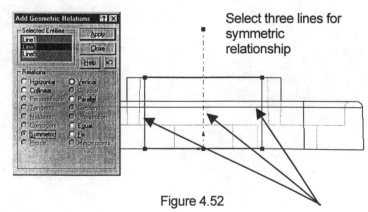

Select three lines for symmetric relationship

Figure 4.52

Click **Apply**, Figure 4.52.  Click **Close**.

Note: You can not select the horizontal lines, Line2 and Line4 if you are creating a vertical symmetric relation.

Dimension the *Sketch*.

Click the **Dimension** icon.  Click the **bottom horizontal line**.  Click the **text location**.  Enter **1.000**, Figure 4.53.

Note: The *Sketch* must be fully defined.

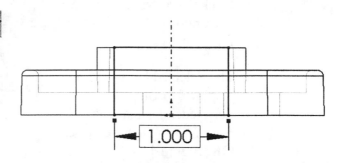

Figure 4.53

Display the *Isometric* view.  Click the **Isometric** icon.

Extrude the *Sketch*.  Click the **Extruded Boss/Base** icon.

Create the first depth direction, Direction 1.

Blind is the Type option.

Enter **0.400** for Depth.

Click the **Draft While Extruding** check box.

Enter **1** for Angle.

Click the **Both Directions** check box, Figure 4.54.

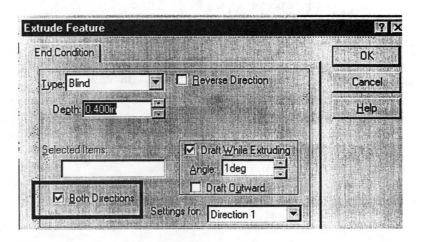

Figure 4.54

Click **Direction 2** from the Setting for list box. Click **Up To Surface** from the Type list box, Figure 4.55.

Click the **outside right rectangular face**, Figure 4.56.

The Selected Items text box displays, "1 Surface Selected." Display the *Boss-Extrude* feature. Click **OK**, Figure 4.57.

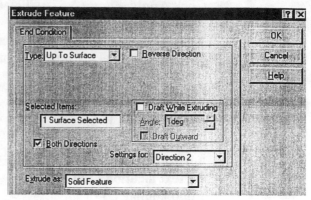

Figure 4.55

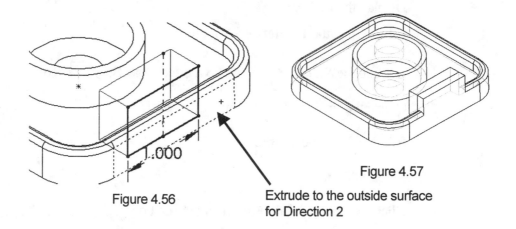

Figure 4.56

Figure 4.57

Extrude to the outside surface
for Direction 2

Rename *Boss-Extrude2* to *ConnectorBase*.

**Save** BATTERYPLATE.

Create the second *Boss-Extrude* feature.

Select the *Sketch* plane. Click the **top narrow face**, Figure 4.58.

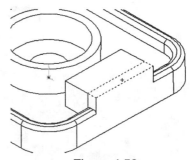

Figure 4.58

Display the *Top* view. Click the **Top** icon, Figure 4.59a.

Sketch the profile. Click the **Sketch** icon. Click the **Centerline** icon. Sketch a **horizontal centerline** with the first point coincident to the

*Origin* .

Create the mirrored centerline. Click **Tools** from the Main menu. Click **Sketch Tools**, **Mirror**. The centerline displays two parallel mirror marks, Figure 4.59b.

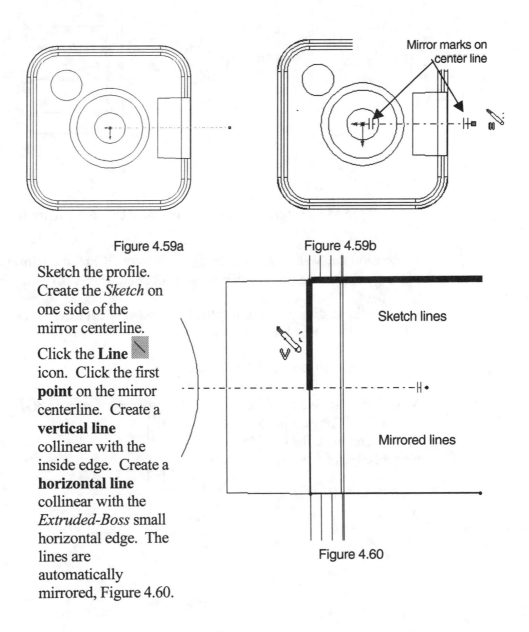

Figure 4.59a                    Figure 4.59b

Sketch the profile. Create the *Sketch* on one side of the mirror centerline.

Click the **Line** icon. Click the first **point** on the mirror centerline. Create a **vertical line** collinear with the inside edge. Create a **horizontal line** collinear with the *Extruded-Boss* small horizontal edge. The lines are automatically mirrored, Figure 4.60.

Mirror marks on center line

Sketch lines

Mirrored lines

Figure 4.60

Complete the *Sketch*. Create a Tangent Arc. Click the **Tangent Arc** icon. Create the first arc point. Click the **endpoint** of the horizontal line. Create a 180 degree arc. Drag the **mouse pointer** downward until the start point, center point and end point are vertically aligned. Release the **mouse button**, Figure 4.61.

Complete the *Sketch*. Click the **Line** icon. Create a **horizontal line**. Create a **vertical line** coincident with the inside top right edge, Figure 4.62.

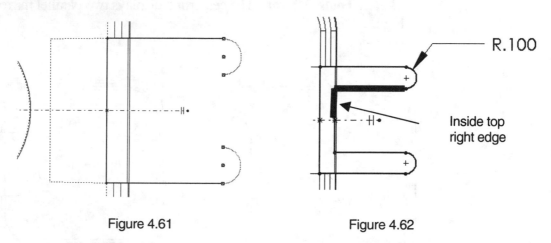

R.100

Inside top
right edge

| Figure 4.61 | Figure 4.62 |

Turn off the Mirror function. Click **Tools, Sketch Tools, Mirror**.

Dimension the *Sketch*. Create a radial dimension. Click the **Dimension** icon. Click the **arc edge**. Click the **text location**. Enter **0.100** for the Radius.

Create a linear dimension. Click the left most **vertical line**. Click the **arc edge**. Click the **text location**, Figure 4.63.

The linear dimension uses the arc center point as a reference. Modify the Properties of the dimension. The Maximize option references the outside tangent edge of the arc.

The sketch is displayed in black.

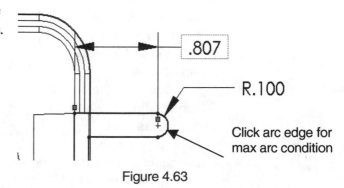

.807

R.100

Click arc edge for
max arc condition

Figure 4.63

Right-click on the **dimension text**. Click **Properties** from the Pop-up menu. Click the **Max** button from the First arc condition option, Figure 4.64. Enter **1.000** in the Value list box. Display the dimension. Click **OK**. The black *Sketch* is fully defined.

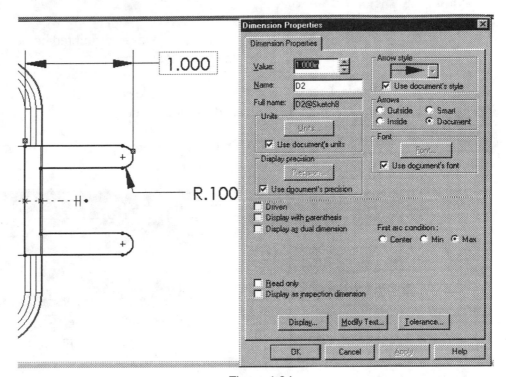

Figure 4.64

Note: Click the arc edge not the arc center point to create a max dimension.

Display the *Isometric* view. Click the **Isometric** icon.

Extrude the *Sketch*. Click the **Extruded Boss/Base** icon. Blind is the default Type option. Enter **0.100** for Depth. Click the **Reverse** check box. Display the *Extrude/Boss* feature. Click **OK**, Figure 4.65.

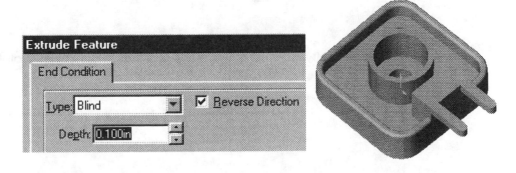

Figure 4.65

Rename ***Boss-Extrude3*** to ***ConnectorSwitch***.

**Save** the BATTERYPLATE.

**4.5.6      Disjoint Geometry**

Incorrect selection of edges and faces leads to disjointed bodies.  Disjointed bodies occur when the geometry contains gaps.

Example: Create a *Sketch* from the outside edge, Figure 4.66.  Reverse the extrusion direction to create disjoint geometry.

The feature is not created and a Rebuild error is displayed, Figure 4.67.

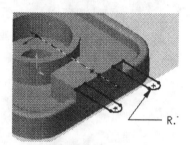

Figure 4.66

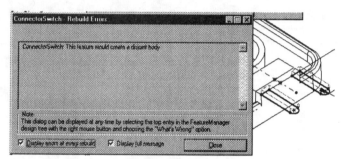

Figure 4.67

Profiles of disjointed and joined geometry are displayed in Figure 4.68.

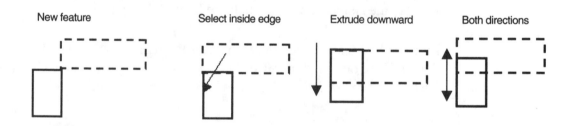

Figure 4.68

### 4.5.7    Create Edge and Face Fillets

Both face and edge options for *Fillet* features are used to smooth rough edges.

Create a *Fillet* on the outside edge of the *Holder*.  Click the **Fillet** ⬛ icon. Click the **outside circular edge** of the *Holder*, Figure 4.69.  Enter **0.100** for Radius.  Display the *Fillet*.  Click **OK**, Figure 4.70.

Rename *Fillet3* to *HolderFillet*.

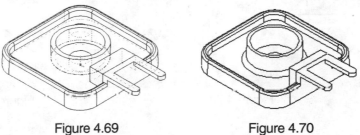

| Figure 4.69 | Figure 4.70 |
|:-----------:|:-----------:|

Create a *Fillet* on the outside bottom edge of the *Connector*.  This is a two step process:

- Create an edge *Fillet*

- *Create a* face *Fillet*

First, create an edge *Fillet* on the two vertical edges of the *Connector*.  Click the **Fillet** ⬛ icon.  Click the two **vertical back edges**, Figure 4.71.

Enter **0.100** for Radius.  Display the *Fillet*.  Click **OK**, Figure 4.72.

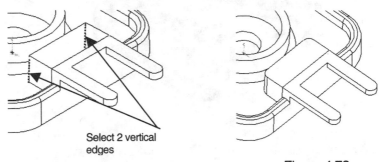

Select 2 vertical edges

| Figure 4.71 | Figure 4.72 |
|:-----------:|:-----------:|

Rename *Fillet4* to *Connect Base Fillet Edge*.

Second, create the face *Fillet*. Click the **Fillet** ⬛ icon. Click **Face Blend** from the Fillet type list box, Figure 4.73. Select the first Face set. Click the **back face**, Figure 4.74.

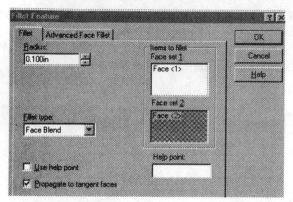

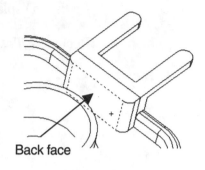

Back face

|  |  |
|---|---|
| Figure 4.73 | Figure 4.74 |

Select the second Face set. Click **inside** the Face set 2 list box. Click the **top face** of the *Base* feature, Figure 4.75. Enter **0.100** for Radius. Click **OK**. Rebuild errors occur.

The Radius is too large. Enter **0.050** for Radius. Display the *Fillet*. Click **OK**, Figure 4.76.

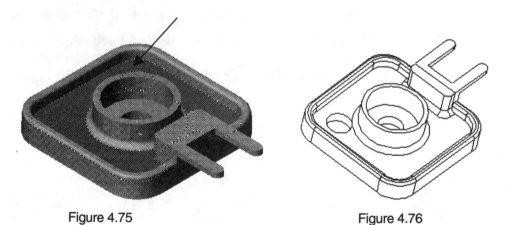

|  |  |
|---|---|
| Figure 4.75 | Figure 4.76 |

Rename *Fillet5* to *Connect Base Fillet Face*.

The FeatureManager displays all feature name icons in yellow, Figure 4.77. The BATTERYPLATE is complete, Figure 4.78.

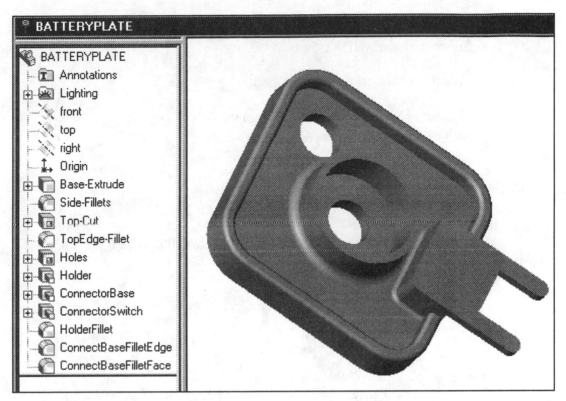

Figure 4.77                                        Figure 4.78

**Save** the BATTERYPLATE.

## 4.6 LENS

The LENS is a purchase part. Obtain dimensional information on the LENS assembly. Review the size, material and construction. Determine the key LENS features.

The *Base* feature for the LENS is a solid *Revolved* feature. A solid *Revolved* feature adds material.

The LENS assembly is comprised of the LENS and BULB. The *Revolved Base* feature is the foundation for the LENS.

A *Revolved* feature is geometry created by rotating a sketched profile around a centerline. Close the *Sketch* profile for a solid *Revolved* feature. Do not cross the centerline.

### 4.6.1   LENS Overview

Create the LENS.  Use the solid *Revolved Base* feature, Figure 4.79.

Create uniform wall thickness.

Create the *Shell* feature, Figure 4.80.

Create an *Extruded-Boss* feature from the back of the LENS, Figure 4.81.

Create a *Thin-Revolved* feature to connect the LENS to the BATTERYPLATE, Figure 4.82.

Figure 4.79

Create a *Counterbore Hole* feature with the *HoleWizard*, Figure 4.83.

Figure 4.80            Figure 4.81            Figure 4.82

The BULB is located inside the *Counterbore Hole*.

Create the front *LensFlange* feature.  Add a transparent *LensShield* feature, Figure 4.84.

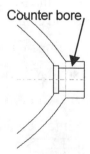

Counter bore

Figure 4.83                          Figure 4.84

### 4.6.2    Create the Revolved Base Feature

Create the LENS.  Click the **New** 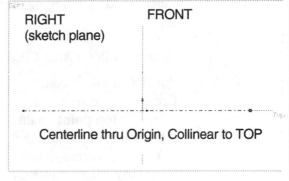 icon.  Click **PartEnglishTemplate**.
Click **OK**.  Click the **Save** icon.  Enter the part name.  Enter **LENS**.  Click **Save**.

The solid *Revolved Base* feature requires a sketched profile and a centerline.
The profile is located on the *Right* plane with the centerline collinear to the
*Top* plane. The profile lines reference the *Top* and *Front* planes.  The curve of
the LENS is created with a
3 point arc.

View the planes.  Right click
on the *Front* plane.  Click
**Show**.  Show the *Top* plane.

Select the *Sketch* plane.  Click
the *Right* plane.

Create the *Sketch*.  Click the

**Sketch** icon.

RIGHT
(sketch plane)

FRONT

Centerline thru Origin, Collinear to TOP

Figure 4.85a

Display the view.  Click the **Right** icon.

Sketch the centerline.  Click the **Centerline** icon.

Sketch a horizontal **centerline** collinear to the *Top* plane, through the *Origin*
, Figure 4.85a.

Sketch the profile.  Create three lines.  Click the **Line** icon.

Create the first line.  Sketch a **vertical line** collinear
to the *Front* plane coincident with the *Origin*.

Create the second line.  Sketch a **horizontal line**
coincident with the *Top* plane.

Create the third line.  Sketch a **vertical line**
approximately 1/3 the length of the first line,
Figure 4.85b.

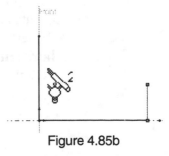

Figure 4.85b

Create an arc.  Determine the curvature of the LENS.

A 3 POINT Arc requires a:

- Start point

- End point

- Center point

The arc midpoint is aligned with the center point.  The arc position is determined by dragging the arc midpoint or center point above or below the arc.

Create a 3 Point Arc.  Click the **3Pt Arc** icon.  Create the arc start point.  Click the **top point** on the left vertical line.  Hold the **left mouse button** down.  Drag the mouse pointer to the **top point** on the right vertical line.

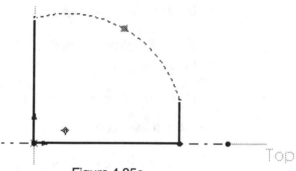

Figure 4.85c

Create the arc end point.  Release the **mouse button**, Figure 4.85c.

Click and drag the **arc center point** below the *Origin*.  Release the **left mouse button**, Figure 4.85d.

Add geometric relationships.

Click the **Add Relations** icon.  Create an equal relationship.  Click the **left vertical line**.  Click the **horizontal line**.  Click the **Equal** button.  Click **Apply**.  Click **Close**.

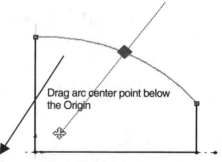

Drag arc center point below the Origin

Figure 4.85d

Add dimensions.  Click the **Dimension**  icon.

Create a vertical linear dimension for the left line.  Enter **2.000**.

Create a vertical linear dimension for the right line.  Enter **0.400**.

Create a radial dimension.  Enter **4.000**, Figure 4.86.  The black *Sketch* is fully defined.

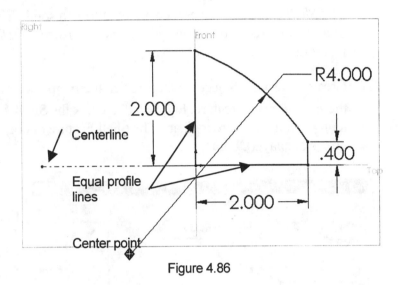

Figure 4.86

Revolve the *Sketch*.  Click the **Revolve** icon from the Feature toolbar.  The Revolve Feature dialog box is displayed, Figure 4.87.

Accept the default option values.  Create the solid *Revolve* feature.  Click **OK**, Figure 4.88.

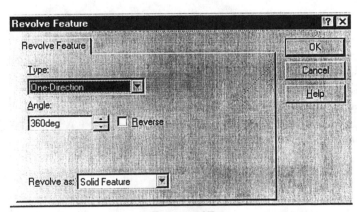

Figure 4.87

Figure 4.88

*Revolve* features contain an axis of revolution. The axis is critical to align other features. Display the axis of revolution. Click **View** from the Main menu. Click **Temporary Axis**. Solid *Revolve* features must contain a closed profile. Each revolved profile requires a individual sketched centerline.

**Save** the LENS.

### 4.6.3    Create a Shell Feature

The *Shell* feature removes face material from a solid. The *Shell* feature requires a face and thickness. Use the *Shell* feature to create thin-walled parts.

Create the *Shell*. Select the face. Click the **front face** of the *Base-Revolve* feature, Figure 4.89. Click the **Shell**  icon from the Feature toolbar. The Shell Feature dialog box is displayed, Figure 4.90.

Figure 4.89

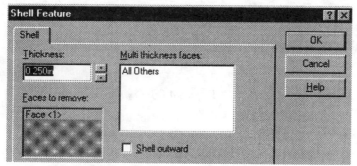

Figure 4.90

Enter **0.250** in the Thickness text box. Display the *Shell* feature. Click **OK**, Figure 4.91.

Rename *Shell1* to *LensShell*.

**Save** the LENS.

Figure 4.91

### 4.6.4    Create the Lens Neck with the Extruded-Boss Feature

Create the *LensNeck*. Use the *Extruded-Boss* feature. The *LensNeck* houses the BULB base and is connected to the BATTERY PLATE. The feature extracts the back circular edge from the *Base-Revolve* feature.

Select the *Sketch* plane. Select the hidden face. Right click near the small **back face**. Click **Select Other** ⌐ from the Pop-up menu. Click the **right mouse button** (N) until the back face is displayed, Figure 4.92a. Accept the back face. Click the **left mouse button** (Y). **Rotate** the part to view the back face, Figure 4.92b.

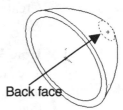

Figure 4.92a

Figure 4.92b

Create the profile. Click the **Sketch** 🖊 icon. Extract the **face** to the *Sketch* plane. Click the **Convert Entities** ▢ icon.

Extrude the *Sketch*. Click the **Extrude Boss/Base** 🖼 icon. Enter **0.400** for Depth. Display the *Extruded-Boss* feature. Click **OK**, Figure 4.93.

Rename *Extrude-Boss* to *LensNeck*.

Hide the Temporary Axis and the Planes. Click **Temporary Axis**. Click **Planes** from the View menu.

**Save** the LENS.

Figure 4.93

### 4.6.5  Create a Counterbore Hole Feature

The LENS requires a *Counterbore Hole* feature. Use the *HoleWizard*. *HoleWizard* assists in creating complex and simple *Hole* features.

Select the *Sketch* plane. Click the **Front** 🔲 icon, Figure 4.94. Click the small **inside back face** of the *Base-Revolve* feature.

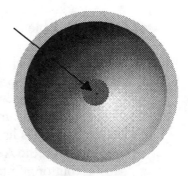

Figure 4.94

Create the *Counterbore Hole*. Click the

**HoleWizard** 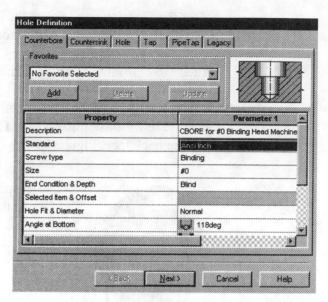 icon. The Hole Definition dialog box is displayed, Figure 4.95a.

Define the own parameters. Click the Parameter 1 **Binding** in the Screw type property text box. The Parameter 1 and Parameter 2 text boxes are displayed, Figure 4.95b.

Enter **Hex Bolt** from the drop down list for Screw type.

Figure 4.95a

Enter ½ from the drop down list for Size.

Click **Through All** from the drop down list for End Condition & Depth.

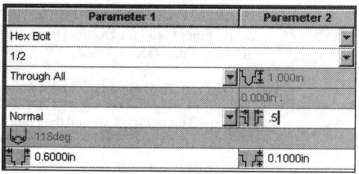

Figure 4.95b

Double-click the drill through **Diameter** value. Enter **.500**.

Double-click the **C-Bore Diameter** value. Enter **0.600**.

Double-click the **C-Bore Depth** value. Enter **0.100**.

Add the new hole type to your favorites list. Click the **Add** button. Enter **CBORE FOR BULB**, Figure 4.95c.

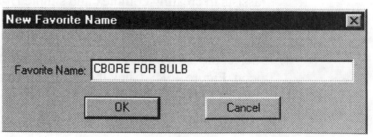

Figure 4.95c

Click **Next** from the Hole Definition dialog box. Position the hole coincident with the *Origin*. Click the **Add Relations**  icon. Click the **center point** of the Counterbore hole. Click the

Origin ⌐. Click **Coincident**. Click **Close**. Complete the hole. Click **Finish** from the Hole Wizard, Figure 4.95d.

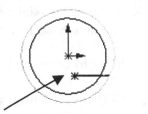

Figure 4.95d

Expand the *Hole*. Click the **Plus Sign** ⊞ icon to the left of the *Hole* feature, Figure 4.96.

*Sketch3* and *Sketch4* are used to create the *Hole* feature.

Rename *Hole1* to ***BulbHole***.

Figure 4.96

Display the *Section* view of *BulbHole* through the *Right* plane. Click the ***Right*** plane from the FeatureManager. Click **View** from the Main menu. Click **Display**, *SectionView*. Click the **Flip Side to View** check box. Click **OK**, Figure 4.97.

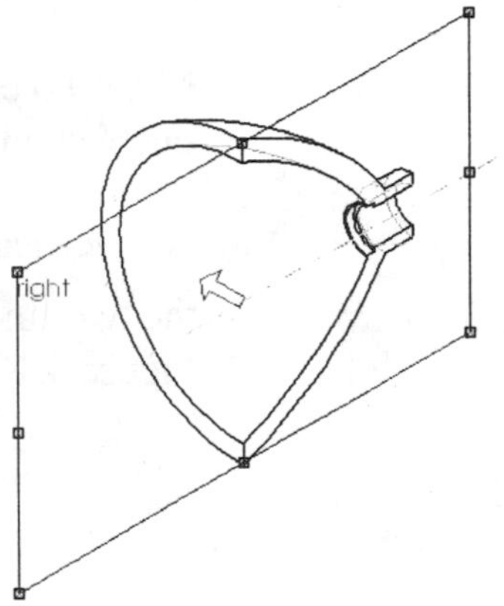

Display the *Full* view. Click **View, Display, SectionView**.

**Save** the LENS.

### 4.6.6 Create the Lens Connector with the Thin-Revolved Feature

Create a *Thin-Revolved* feature. Rotate a sketch open profile around a centerline. Note: The sketch profile must be open and cannot cross the centerline.

Figure 4.97

Use the *Thin-Revolved* feature to connect the LENS to the BATTERYPLATE.

Select the *Sketch* plane. Click the ***Right*** plane. Display the *Right* view. Click the **Right** icon.

Create the *Sketch*. Click the **Sketch** icon.

Sketch the centerline. Click the **Centerline** icon. Sketch a **horizontal centerline** collinear to the *Top* plane through the *Origin* ⌐.

Sketch the profile. **Right-click** in the Graphics window. Click **Select** from the Pop-up memu.

Click the **right edge** of the *Base* feature.

Click the **Convert Entities** icon. Extract the edge, Figure 4.98. Create an arc tangent to the extracted edge.

Click the **TangentArc** icon. Click the **top point** of the vertical line. Drag the **mouse pointer** to the left.

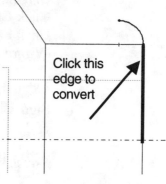

Click this edge to convert

Figure 4.98

The mouse pointer displays a vertical line when the endpoint aligns with the arc center point, Figure 4.99. Create the 90-degree arc. Release the **left mouse button**.

Note: To create a 90-degree arc, the Snap to points in the Grid/Units must be unchecked.

The vertical line segments were required to create the Tangent Arc. Remove the two line segments. Click the **Trim** icon. Click **both vertical edges**, Figure 4.100. The *Sketch* consists of an arc and a centerline.

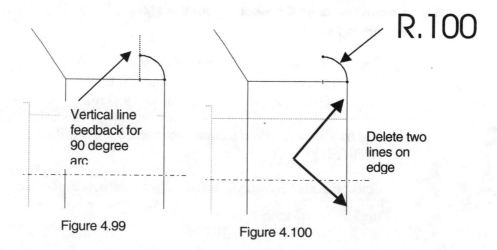

Vertical line feedback for 90 degree arc.

R.100

Delete two lines on edge

Figure 4.99                    Figure 4.100

Add dimensions. Click the **Dimension** icon. Create a radial dimension. Enter **0.100**.

The sketch arc requires a coincident relationship. This insures that the center point of the arc is coincident with the horizontal silhouette edge of the *Base-Revolve* feature, Figure 4.101.

Add geometric relations. Click the **Add Relations** icon. Click the **arc center point**. Click the **horizontal line** (silhouette edge) of the *Base-Revolve* feature. Click the **Coincident** button. Click **Apply**. Click **Close**.

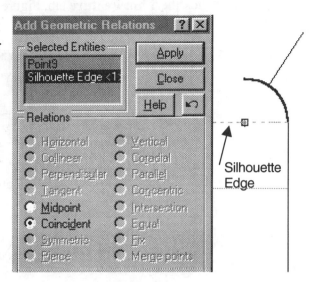

Figure 4.101

The black *Sketch* is fully defined.

Revolve the *Sketch*. Click the **Revolve** icon. A warning message appears, Figure 4.102. Keep the *Sketch* open. Click **No**.

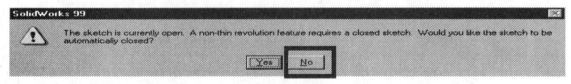

Figure 4.102

The Revolve Thin Feature dialog box is displayed, Figure 4.103.

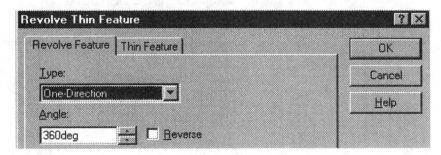

Figure 4.103

Click the **Thin Feature** tab, Figure 4.104.

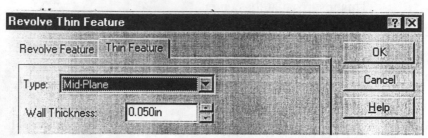

Figure 4.104

Create the *Thin-Revolved* feature on both sides of the *Sketch*. Select **Mid-Plane** from the Type list box. Enter **0.050** for Wall Thickness, Figure 4.105.

Click **OK** from the Revolve Thin Feature dialog box, Figure 4.106.

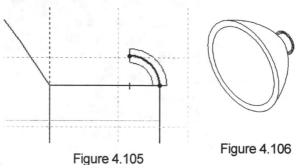

Figure 4.105

Figure 4.106

Rename ***Boss-Revolve-Thin*** to ***LensConnector***.

**Save** the LENS.

### 4.6.7  Create a Lens Cover with an Extruded-Boss Feature

Use the *Extruded-Boss* feature to create the front *LensCover*. The feature extracts the front inside circular edge from the *Base-Revolve* feature. The front *LensCover* is a key feature for designing the mating component. The mating component is the LENSCAP.

Select the *Sketch* plane. Click the **front circular face**, Figure 4.107a.

Figure 4.107a

Create the *Sketch*. Click the **Sketch** icon.
Display the *Front* view. Click the **Front** icon.
Click the **inside circular edge**, Figure 4.107b.
Click the **Convert Entities** icon.

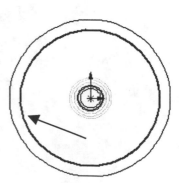

Figure 4.107b

Click the resultant inside **circle**. Click the **Offset Entities** icon. Enter **0.500**. Click **Apply, Close**, Figure 4.107c.

Note: Do not select the outside edge. You will create a disjointed feature.

Extrude the *Sketch*. Click the **Extrude Boss/Base** icon. Enter **0.250** for Depth.

Display the LENS. Click **OK**, Figure 4.107d.

Rename *Boss-Extrude* to *LensCover*.

**Save** the LENS.

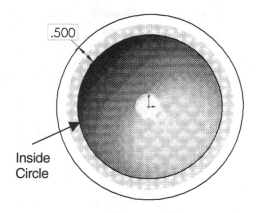

Figure 4.107c

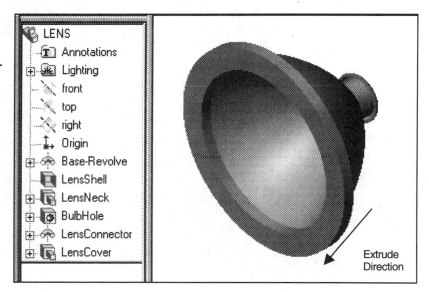

Figure 4.107d

**4.6.8      Create the Lens Shield with an Extruded-Boss Feature**

An *Extruded-Boss* feature is used to create the front *LensShield*. The feature extracts the front inside circulcar edge from the *Base-Revolve* feature. The *LensShield* front and back circular faces are transparent in order to view the BULB and simulate clear plastic.

Select the *Sketch* plane. Click the **Front** plane, Figure 4.108a.

Sketch the profile. Click the **Sketch** 🖉 icon. Display the *Front* view. Click the **Front** 🔲 icon. Click the **inner circular edge**, Figure 4.108b. Click the **Convert Entities** 🗗 icon. The circle is projected onto the *Front* Plane.

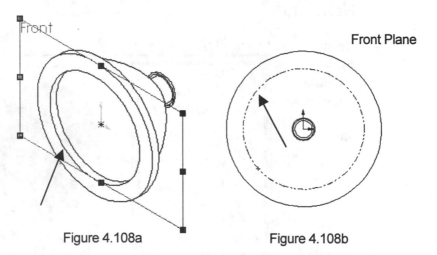

Figure 4.108a                    Figure 4.108b

Extrude the Sketch. Click the **Extruded Boss/Base** 🔲 icon. Enter **0.100** for Depth. Click **OK**, Figure 4.108c.

Rename **Boss-Extrude3** to *LensShield*.

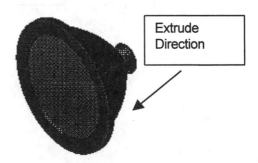

Figure 4.108c

Add transparency to the front and back circular faces of the *LensShield*. Right-click the *LensShield* in the Graphics window, Figure 4.108d.

Click **Feature Properties**. The Feature Properties dialog box is displayed, Figure 4.108e. Click the **Color** button. The Entity Property dialog box is displayed, Figure 4.108f. Click the **Advanced** button.

Set the transparency for the faces. Drag the **Transparency slider** to the far right side, Figure 4.108g.

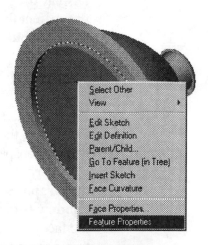

Figure 4.108d

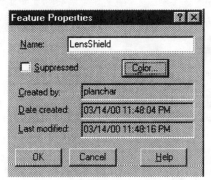

Figure 4.108e

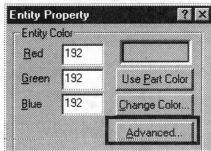

Figure 4.108f

Click **OK** from the Material Properties dialog box. Click **OK** from the Entity Property dialog box.

Click **OK** from the Feature Properties box.

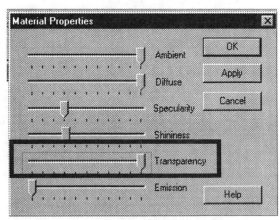

Figure 4.108g

Display the
transparent faces.
Click the **Shaded**
icon,
Figure 4.108h.

When the *LensShield*
is selected, the faces
are not transparent.
Click anywhere in
the **Graphics**
**window** to display
the face
transparency.

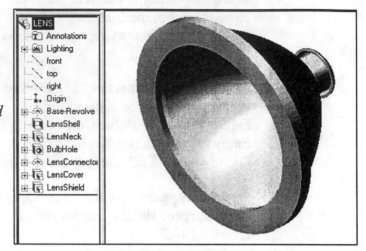

Figure 4.108h

**Save** the LENS.

## 4.7 BULB

The BULB is contained within the LENS assembly. The BULB is a
purchased part. The BULB utilizes the *Revolved* feature as the *Base* feature.

### 4.7.1 BULB Overview

- Create the *Revolved-Base* feature from a sketched profile on the *Right*
  plane, Figure 4.109a.

- Create a *Revolved* feature using a B-Spline sketched profile. A B-Spline is
  a complex curve, Figure 4.109b.

- Create a *Revolved Cut* feature at the base of the BULB, Figure 4.109c.

- Create a *Dome* feature at the base of the BULB, Figure 4.109d.

- Create a *Circular Pattern* feature from an *Extruded-Cut*, Figure 4.109e.

Figure 4.109a          4.109b          4.109c          4.109d          4.109e.

### 4.7.2    Create the Revolved Base Feature

Create the BULB.  Click the **New** ☐ icon.  Click **PartEnglishTemplate**.
Click **OK**.  Click the **Save** 🖫 icon.  Enter the part name.  Enter **BULB**.
Click **Save**.

The solid *Revolved-Base* feature requires a centerline and a sketched profile.

Select the *Sketch* plane.  Click the **Right** plane.

Sketch the centerline.  Click the **Sketch** ✎ icon.  Display the *Right* view.
Click the **Right** 🔲 icon.  Click the **Centerline** ┆ icon.  Sketch a horizontal
**centerline** collinear to the *Top* plane through the *Origin* ⌁.

The flange of the BULB is located inside the *Counterbore Hole* of the LENS.
Align the bottom of the flange with the *Front* plane.  The *Front* plane mates
against the *Counterbore* face.

Right-click the **Front** plane in the FeatureManager.
Click **Show**.  Sketch the profile.  Create six lines.
Click the **Line** ◼ icon.

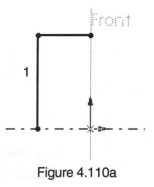

Figure 4.110a

1.  Create the first line.  Sketch a **vertical line** to
    the left of the *Front* plane, Figure 4.110a.

2.  Create the second line.  Sketch a **horizontal
    line** with the endpoint coincident to the *Front*
    plane.

3.  Create the third line.  Sketch a short
    **vertical line** towards the centerline,
    collinear with the *Front* plane,
    Figure 4.110b.

4.  Create the forth line.  Sketch a **horizontal
    line** to the right.

5.  Create the fifth line.  Sketch a **vertical line**
    with the endpoint collinear with the
    centerline.

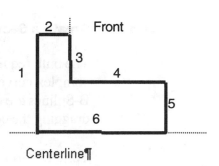

Figure 4.110b

6.  Create the sixth line.  Close the *Sketch*.
    Sketch a **horizontal line**.

Add dimensions, Figure 4.110c. Click the **Dimension** ✎ icon.

- Create a vertical linear dimension. Click the **right line**. Enter **.200**.

- Create a vertical linear dimension. Click the **left line**. Enter **.295**.

- Create a horizontal linear dimension. Click the **top left line**. Enter **0.100**.

- Create a horizontal linear dimension. Click the **top right line**. Enter **0.500**.

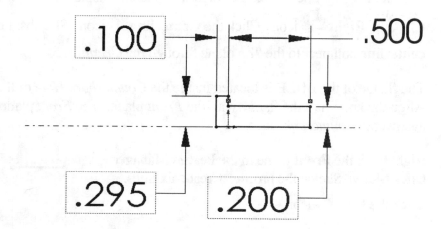

Figure 4.110c

Revolve the *Sketch*. Click the **Revolve** 🦅 icon from the Feature toolbar. The Revolve Feature dialog box is displayed. Accept the default option values. Click **OK**, Figure 4.111.

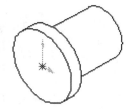

Save the BULB.

Figure 4.111

### 4.7.3    Create a Second Revolve Feature

The bulb requires a second solid *Revolve* feature. The profile utilizes a complex curve called a B-Spline (Non-Uniform Rational B-Spline or NURB). B-Splines are drawn with control points. Adjust the shape of the curve by dragging the control points.

Turn the Grid Snap off. Click the **Grid** ▦ icon. Uncheck the **Snap to points** check box.

Select the *Sketch* plane. Click the ***Right*** plane.

Create the Sketch. Click the **Sketch** icon. Display the Right view. Click the **Right** icon.

Sketch the centerline. Click the **Centerline** icon. Sketch a **horizontal centerline** collinear to the *Top* plane, coincident to the *Origin*.

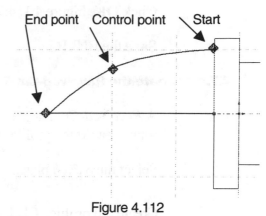

End point    Control point    Start

Figure 4.112

Sketch the profile. Click the **B-Spline** icon. Sketch the start point. Click the **left vertical edge** of the *Base* feature, Figure 4.112.

Sketch the control point. Drag the **mouse pointer** to the left of the *Base* feature and below the first point. Release the **left mouse** button.

Sketch the end point. Click the **control point**. **Drag** the mouse pointer to the center line. Release the **left mouse button**.

Adjust the B-Spline. Click the **Select** icon. Position the **mouse pointer** over the B-Spline control point. Drag the **mouse pointer** upward. Release the **left mouse button**.

Note: SolidWorks does not require dimensions to create a feature.

Complete the profile. Sketch two lines. Click the **Line** icon.

Create a horizontal line. Sketch a **horizontal line** from the B-Spline endpoint to the left edge of the *Base-Revolved* feature.

Create a vertical line. Sketch a **vertical line** to the B-Spline start point, collinear with the left edge of the Base-Revolved feature, Figure 4.113.

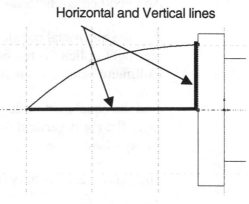

Horizontal and Vertical lines

Figure 4.113

Revolve the *Sketch*. Click the **Revolve** 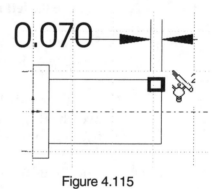 icon from the Feature toolbar. The Revolve Feature dialog box is displayed. Accept the default option. Display the Revolve feature. Click **OK**, Figure 4.114.

**Save** the BULB.

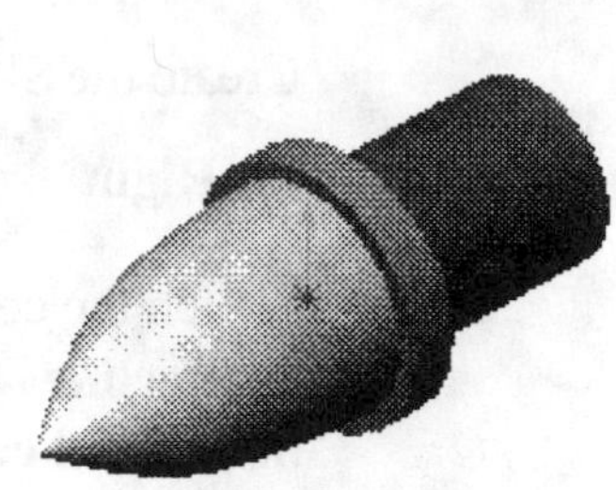

### 4.7.4 Create the Revolved-Cut Feature

A *Revolved-Cut* feature removes material by rotating a sketch profile around a centerline.

Select the *Sketch* plane. Click the ***Right*** plane.

Create the profile. Click the **Sketch** icon. Display the *Right* view. Click the **Right** icon.

Sketch the centerline. Click the **Centerline** icon. Sketch a **horizontal centerline** collinear to the *Top* plane, coincident to the *Origin*.

Sketch the profile. Click the **Rectangle** icon. Sketch a **rectangle** with the first point coincident with the top right corner of the *Base* feature, Figure 4.115.

Add dimensions. Click the **Dimension** icon. Create the top horizontal dimension. Enter **0.070**.

Add geometric relations. Click the **Add Relations** icon.

Add the horizontal collinear relationship. Click the **top horizontal line** of the rectangle. Click the **top edge** of the *Base-Extrude* feature. Click the **Collinear** button. Click **Apply**.

Add the equal relationship. Click the **top horizontal line** of the rectangle. Click the **right vertical line** of the rectangle. Click the **Equal** button. Click **Apply**. Click **Close**.

The black *Sketch* is fully defined.

Figure 4.114

Figure 4.115

Revolve the *Sketch*. Click the **Revolved**

**Cut**   icon from the Feature toolbar. The Revolve Feature dialog box is displayed. Accept the default option values.

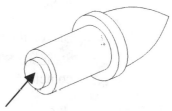

Figure 4.116

Display the *Cut* feature. Click **OK**, Figure 4.116.

**Save** the BULB.

### 4.7.5  Create the Dome Feature

A *Dome* feature creates spherical or elliptical shaped geometry. Use the *Dome* feature to create the *Connector* feature of the BULB.

Figure 4.117

Create the *Dome* feature. Select the *Sketch* plane. Click the **back circular face** of the *Revolve Cut,* Figure 4.117.

Click **Insert** from the Main menu. Click **Features**, **Dome**. The Dome dialog box is displayed, Figure 4.118.

Enter **0.100** for Height. Display the *Dome*. Click **OK**, Figure 4.119.

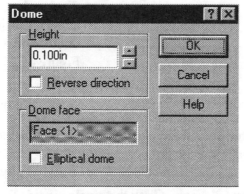

Figure 4.118

Figure 4.119

**Save** the BULB.

### 4.7.6  Create a Circular Pattern

The *Pattern* feature creates one or more instances of a feature or a group of features. The *Circular Pattern* feature places the instances around an axis of revolution.

The *Pattern* feature requires a seed feature. The seed feature is the first feature in the *Pattern*. The seed feature in this section is an *Extruded-Cut*.

Select the *Sketch* plane. Click the **front circular face** of the *Base* feature, Figure 4.120.

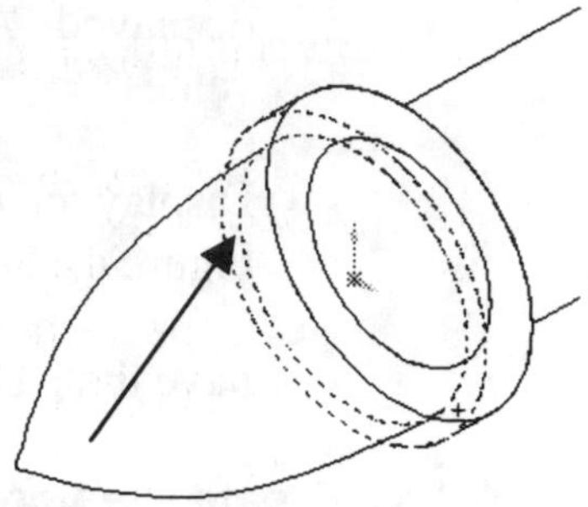

Create the *Sketch*. Click the **Sketch** icon. Display the *Front* view. Click the **Front** icon.

Sketch the centerline. Click the **Centerline** icon. Sketch a **vertical centerline** coincident with the top and bottom circular circles, Figure 4.121a.

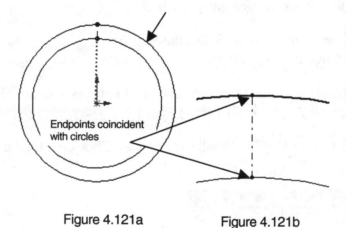

Endpoints coincident with circles

Figure 4.120

Extract the outside circular edge. Extract the outside circular edge. Click the **Select** icon. Click the **outside circular edge**. Click the **Convert Entities** icon. **Zoom** to display the centerline and the outside circular edge, Figure 4.121b.

Figure 4.121a                    Figure 4.121b

Sketch a V-shaped line. Click the **Line** icon. Create the first point. Click the **midpoint** of the centerline. Create the second point. Click the coincident **outside circle edge**, Figure 4.122.

Trim the lines. Click the **Trim** icon. Click the **circle** outside the V shape, Figure 4.123.

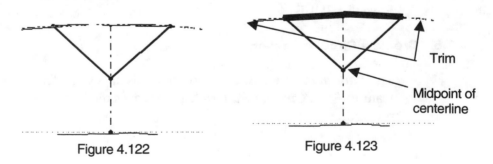

Trim

Midpoint of centerline

Figure 4.122                              Figure 4.123

Add the geometry relations. Click the **Add Relations** ⊥ icon. Add the perpendicular relations. Click the **two lines**. Click the **Perpendicular** button. Click **Apply**. Click **Close**.

The black *Sketch* is fully defined.

Extrude the *Sketch*. Click the **Extruded Cut** 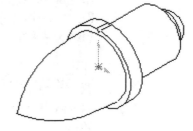 icon. Click **Up to Next** from the Type list box. Display the *Extruded-Cut*. Click **OK**, Figure 4.124.

Figure 4.124

Display the Temporary axis. Click **View**, **Temporary Axis** from the Main menu. Click the **Direction selected** text box. Click **Temporary Axis**.

The *Cut-Extrude* is the seed feature for the *Pattern*.

Create the *Pattern*. Click the **Cut-Extrude** feature. Click the **Circular Pattern** icon. The Circular Pattern dialog box is displayed, Figure 4.125.

Create 4 copies of the *Cut*.

Enter **180** in the Total angle spin box. Enter **4** in the Total instances spin box. Click the **Equal spacing** check box. Click the **Geometry pattern** check box.

Click **Preview**. Click the **Total angle** Spin box. Increase 180-degree to **360**. The Pattern dynamically changes.

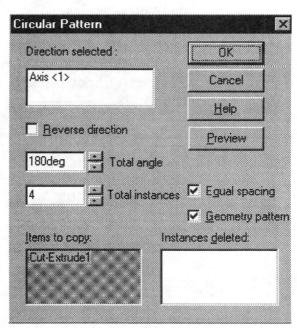

Figure 4.125

Display the *Pattern* feature.  Click **OK**, Figure 4.126.

Edit the *Pattern* feature.  Right-click on the ***Circular Pattern*** from the Feature Manager.  Click **Edit Definition**.  Enter **360** in the Total angle spin box.  Enter **8** in the Total instances spin box.  Display the updated *Pattern*.  Click **OK**, Figure 4.127.

Hide the Temporary axis.  Click **View** from the Main menu.  Click **Temporary Axis**.

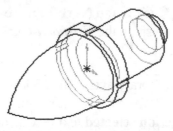 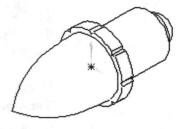

Figure 4.126                    Figure 4.127

**Save** the BULB.

## 4.8   Customizing Toolbars

The default Toolbars contains numerous icons that represent basic functions.  SolidWorks contains additional features and functions not displayed on the default Toolbars.

Place the Mirror icon on the Sketch Tools Toolbar.  Click **Tools** from the Main menu.  Click **Customize**.  The Customize dialog box is displayed, Figure 4.128.

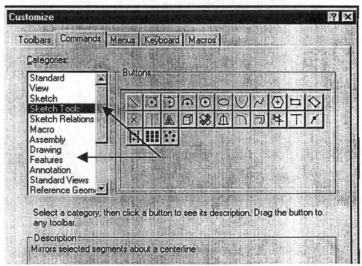

Figure 4.128

Click the **Commands** tab. Click **Sketch Tools** from the category text box.

Click the **Mirror** 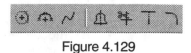 icon. The icon command description is displayed in the Description text box.

Drag the **Mirror** icon into the Sketch Tools Toolbar, Figure 4.129.

Figure 4.129

Place the **Dome** icon on the Features Toolbar.

Click **Features** from the category text box, Figure 4.130. Drag the **Dome** icon into the Features Toolbar, Figure 4.131.

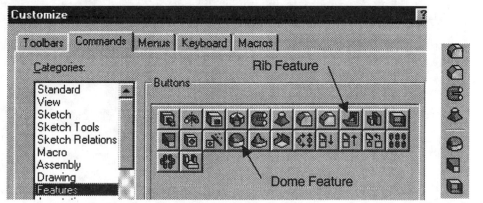

Figure 4.130                                          Figure 4.131

Drag the **Rib** icon into the Features Toolbar.

Update the Toolbars. Click **OK** from the Customize dialog box.

You have just created four parts:

- BATTERY

- BATTERY PLATE

- LENS

- BULB

Practice the exercises before moving onto the next section.

**4.9 Questions**

1.      Identify the function of the following features:

   - Fillet

   - Extruded Cut

   - Boss Extrude

2.      How do you add symmetric relations?

3.      How do you avoid the Fillet Rebuild error?

4.      How do you create an angular dimension?

5.      What is a draft angle?

6.      When do you use a draft angle?

7.      When do you use the Mirror command?

8.      What is disjointed geometry?

9.      What is the function of the Shell feature?

10.     An arc requires _____ points?

11.     Name the required points of an arc?

12.     Why would you use the Hole Wizard feature?

13.     What is a B-Spline?

14.     Identify the required information for a *Circular Pattern*?

15.     How do you add the *Dome* feature icon to the Feature Toolbar?

### 4.10 Exercises

#### I. Create the following Extruded Parts:

Exercise 4.1          MOUNTING PLATE.

Exercise 4.2          L-BRACKET WITH ANGLE SUPPORT.

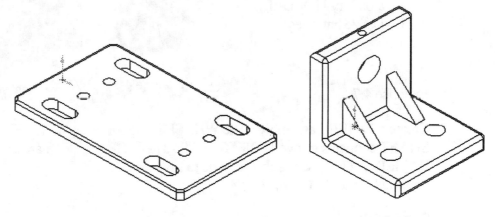

Exercise 4.1                              Exercise 4.2

#### II. Create the following Revolved Parts:

Exercise 4.3          SIMPLE SCREW.

Exercise 4.4          SIMPLE CAP SCREW.

Exercise 4.5          SPOOL.

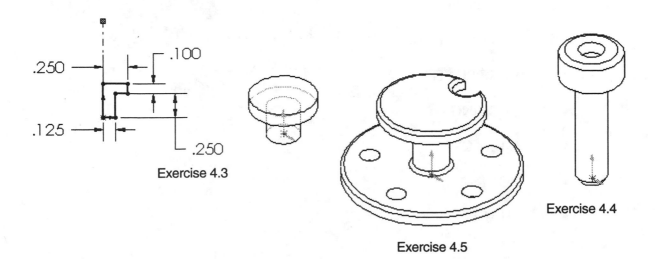

Exercise 4.3

Exercise 4.4

Exercise 4.5

### III. Design Projects.

Exercise 4.6a

Create a D-size battery.

Exercise 4.6b

Create a battery HOLDER to
hold 4-D size batteries.

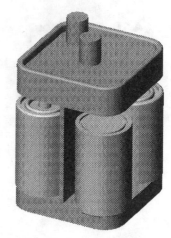

Exercise 4.6a                    Exercise 4.6b

Exercise 4.7

Create a WHEEL assembly. The WHEEL is supported by a SHAFT. The
SHAFT connects to two L-BRACKETS. The L-BRACKETS are mounted to a
BASE PLATE. Use purchased parts to save time and cost. The only dimension
provided is the WHEEL.

Select a WHEEL diameter:

- 3"

- 4"

- 100 mm

Find a material supplier using
the WWW. (Hint:
www.thomasregister.com)

WHEEL Assembly Parts:

- BASE PLATE

- BUSHINGS

- L-BRACKET

- BOLTS

- SHAFT

Exercise 4.7

Exercise 4.8

Create a TRAY and GLASS. Use real objects to determine the overall size and shape of the *Base* feature. Below are a few examples.

Exercise 4.8

Exercise 4.9

Create a JAR-BASE. Save the JAR-BASE as a new part, JAR COVER. Use the dimensions from the JAR-BASE to determine the size of the JAR-COVER.

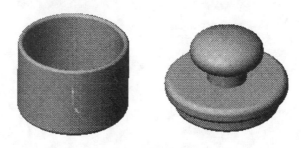

Exercise 4.9

Exercise 4.10

Create an EMBOSSED-STAMP with your initials.

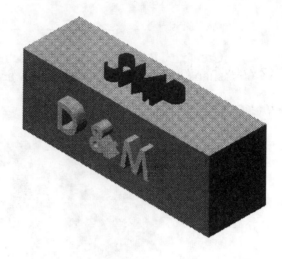

The initials are created with Extruded Sketched text. How do you create the text? Answer: Explore the command with SolidWorks on-line Help. Click the **Help** ? icon. Click **Index**. Enter **text**. Click **extruded text on model**. Follow the instructions.

Exercise 4.10

# Project 5

## Sweep and Loft Features

Below are the desired outcomes and usage competencies based upon the
completion of Project 5.

| Project Desired Outcomes | Usage Competencies |
|---|---|
| • Four key flashlight components: O-RING, SWITCH, LENSCAP and HOUSING. | • Comprehension of the fundamental definitions and process of Feature-Based 3D Solid Modeling using Sweeps and Lofts. |
| • FLASHLIGHT assembly. | • Knowledge to create Sweep and Loft features. |
| | • Ability to combine multiple features to create components. |
| | • Knowledge of assembly Mating functionality. |

## 5    Project 5 – Sweep and Loft Features

### 5.1    Project Objective

Create four components of the FLASHLIGHT assembly.  Create an O-RING, SWITCH, LENSCAP and HOUSING.  Create the final FLASHLIGHT assembly.

### 5.2    Project Situation

Communications is a major component to a successful development program. Provide frequent status reports to the customer and to your team members.

Communicate with suppliers.  Ask questions.  Check on details.  What is the delivery time for the BATTERY, LENS and SWITCH?

Talk to colleagues to obtain useful manufacturing suggestions and ideas.

Your team decided that plastic injection molding is the most cost effective method to produce large quantities of the desired part.

Investigate surface finishes that support the customers advertising requirement? You have two fundamental choices:

- Adhesive label

- Silk-screen

There are time, quantity and cost design constraints.  For the prototype, use an adhesive label.  In the final product, create a silk screen.

Investigate your options on O-Ring material.  Common O-Ring materials for this application are Buna-N (Nitrile®) and Viton®.

Note: Be cognizant of compatibility issues between O-Ring materials and lubricants.

### 5.3    Project Overview

You will create four parts in this project:

- O-RING

- SWITCH

- LENSCAP

- HOUSING

Two *Base* feature types are discussed in this project:

- *Sweep* – O-RING

- *Loft* – SWITCH

The LENSCAP and HOUSING are created with a
combination of features.

Figure 5.1a

Use the *Sweep* feature as the O-RING *Base* feature,
Figure 5.1a. Create the O-RING *Sweep* feature by moving a
small circular profile along a large circular path. The
diameter of the large circular path of the O-RING is
determined by the diameter of the LENS. Position the
O-RING between the LENS and the LENSCAP.

The SWITCH is a purchased part. The SWITCH is a
complex assembly. Use the *Loft* feature to create the
SWITCH, Figure 5.1b.

Figure 5.1b

The LENSCAP and HOUSING are designed plastic parts.
The LENS and the BATTERY is enclosed by the LENSCAP and HOUSING
respectively. How do you design the LENSCAP and HOUSING to ease the
transition from development to manufacturing? Answer: Review the fundamental
design rules behind the plastic injection manufacturing process:

- Maintain a constant wall thickness. Inconsistent wall thickness creates stress.

- Create a radius on all corners. No sharp edges. Sharp edges create vacuum
  issues when removing the mold.

- Allow a minimum draft angle of 1 degree. Draft sides and internal ribs. Draft
  angles assist in removing the part from the mold.

Obtain additional information on material and manufacturing from material
suppliers or manufacturers. Example: GE Plastics of Pittsfield, MA
(www.geplastics.com & www.gepolymerland.com ) provides design guidelines
for selecting raw materials and creating plastic parts and components.

Review the three developed parts in Project 4:

- BATTERY

- LENS

- BATTERYPLATE

The BATTERY and LENS are purchase parts. The BATTERYPLATE is a created part.

Position the BATTERY next to the LENS, Figure 5.2a. The diameter of the LENSCAP must be slightly larger than the diameter of the LENS.

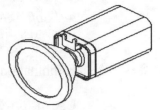

The overall height of the LENSCAP and HOUSING is approximately 7.0". The dimension of the LENS and the BATTERY determines the size of the HOUSING.

Figure 5.2a

Create the LENSCAP. Use the *Base-Extrude* feature with a draft angle, Figure 5.2b.

Create an internal thread in the LENSCAP. Use the *Sweep* feature and a sketched *Helix/Spiral* curve. Revolve a *Boss* feature around the outside of the LENSCAP. Use a *Circular Pattern* feature.

Create the HOUSING. The *Extruded-Base* feature dimensions are determined from the LENS and LENSCAP. The *Loft* feature is used to transition a circular profile of the LENSCAP to a square profile of the BATTERY.

Figure 5.2b

Use the *Extrude* feature to enclose the BATTERY. Create a constant wall thickness with the *Shell* feature. Create the handle. Use the *Sweep* feature.

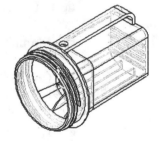

The *Rib* feature provides extra internal support and rigidity, Figure 5.2c. Ribs will maintain the proper alignment of the BATTERY and LENS.

Figure 5.2c

Note: In this project, wall thickness, clearance fit and thread size are increased for improve illustration. Parts are simplified.

Create the final FLASHLIGHT assembly, Figure 5.2d. The FLASHLIGHT assembly consists of the following:

- BATTTERYANDPLATE sub-assembly

- LENSANDBULB sub-assembly

- CAPANDLENS sub-assembly

- SWITCH and HOUSING components

Figure 5.2d

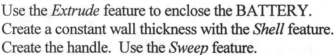

Do not assemble all components at the top level. Create sub-assemblies which represent major pieces of the assembly.

Record the file location of the parts. The assembly references part file locations. Use the Save As and Save as Copy commands within SolidWorks. Do not move parts to a different directory through the Windows File Manager.

## 5.4 O-RING

The O-RING captures the LENS to the LENSCAP. Create the O-RING. Use the *Sweep* feature as the *Base* feature. A *Sweep* feature adds material by moving a profile along a path.

Note: *Sweep* features require a cross section and a path. Example: Think of a doughnut with a bite, Figure 5.3. The cross section is the small circle, "bite" of the doughnut.

Figure 5.3

The path is the large circle that constitutes the circumference of the donut.

This is your O-RING.

### 5.4.1 O-RING Overview

Create the O-RING. The default orientation of the O-RING is based on the orientation in the assembly.

Sketch a large circular path on the *Front* plane. Sketch the small cross section on the *Right* plane, Figure 5.4a. The path and the cross section are combined to create the *Sweep* feature, Figure 5.4b.

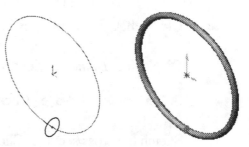

Figure 5.4a            Figure 5.4b

The diameter of the large circular path is 4.35". The outside diameter of the *LensCover* feature is 4.466", Figure 5.4c.

Let's start.

### 5.4.2 Create a Sweep-Base Feature

Create the O-RING. Click the ▢ icon.

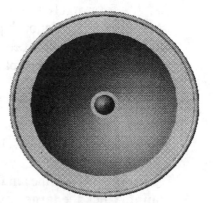

Figure 5.4c

Click **PartEnglishTemplate**. Click **OK**. Click the **Save** ▣ icon. Enter part name. Enter **O-RING**. Click **Save**.

The *Sweep-Base* feature uses:

- A circular path sketched on the *Front* plane

- A small cross section sketched on the *Right* plane

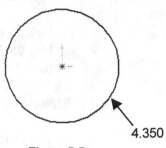

Create the *Sweep* path.

Select the *Sketch* plane. The *Front* plane is the default *Sketch* plane.

Figure 5.5a

Sketch the path. Click the **Sketch**  icon. Click the

**Circle** ⊕ icon. Sketch a circle centered at the *Origin* ↥, Figure 5.5a.

Add the dimensions. Click the **Dimension** icon. Dimension the circumference. Enter **4.350**.

Close the *Sketch*. Click the **Sketch** icon.

Rename *Sketch1* to *sketchpath*.

Create the *Sweep* cross section.

Display the *Isometric* view. Click the **Isometric** icon.

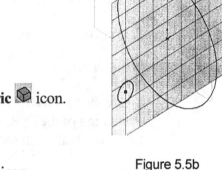

Select the *Sketch* plane. Click the *Right* plane.

Sketch the cross section. Click the **Sketch** icon.

Figure 5.5b

Click the **Circle** ⊕ icon. Create a **small circle** left of the *sketchpath*, Figure 5.5b.

Add geometric relationships. The Pierce function positions the center of the cross section on the sketched path.

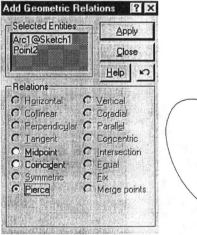

Click the **Add Relations** ⊥ icon. **Delete** the small Arc2 entry from the Selected Entities text box. Right-click in the **Graphics window**. Click **Clear Selections**. Click the **small circle center point**. Click the **large circle circumference**. Click the **Pierce** button, Figure 5.6a.

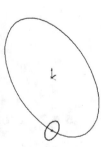

Figure 5.6a            Figure 5.6b

Display the *Sketch*. Click **Apply**. Click **Close**, Figure 5.6b.

Add dimensions. Click the **Dimension** 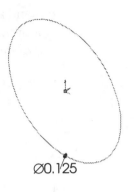 icon.
Dimension the small circle. Enter diameter. Enter **0.125**,
Figure 5.7.

Close the *Sketch*. Click the **Sketch** icon.

Rename *Sketch2* to *sketch-section*.

Note: For detailed cross sections, perform the following:

- Create a large cross section

- Pierce the section to the path

- Add dimensions to create the true size

Ø0.125

Figure 5.7

Sweep the *Sketch*. Click the **Sweep** icon from the Feature toolbar. The Sweep
dialog box is displayed, Figure 5.8.

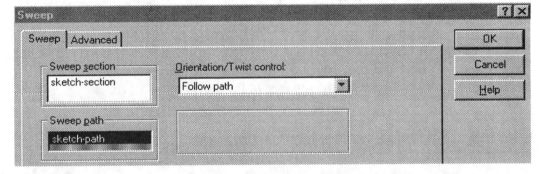

Figure 5.8

Select the Sweep section.
Click the **Sweep section**
text box. Click the
*sketch-section* from the
FeatureManager. Click
the **Sweep path** text box.
Click the *sketch-path*
from the FeatureManager.
Display the Sweep. Click
**OK**, Figure 5.9.

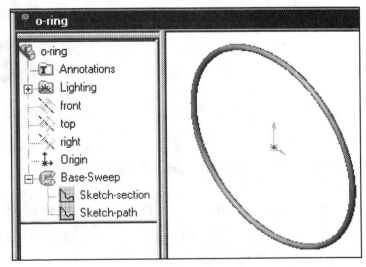

Figure 5.9

Expand the *Base-Sweep* feature. Click the **Plus Sign** ⊞ icon in the FeatureManager. The *Base-Sweep* is composed of the *Sketch-section* and the *Sketch-path*. Collapse the *Base-Sweep* feature. Click the **Minus Sign** ⊟ icon.

**Save** the O-RING.

## 5.5   SWITCH

The SWITCH is a purchased part. The SWITCH is a complex assembly. Create the outside casing. Create the SWITCH with the *Base-Loft* feature, Figure 5.10.

A *Loft* feature uses two or more cross sections to define geometry contours. Sketch each cross section on a plane. The path connects the cross sections to create the solid *Loft* feature.

Figure 5.10

### 5.5.1   SWITCH Overview

The orientation of the SWITCH is based on its position in the assembly. The SWITCH is comprised of three cross sections. Each cross section is sketched on a different plane, Figure 5.11a. Create the *Base-Loft* feature, Figure 5.11b. Create a *Dome* feature to the top face of the *Loft*, Figure 5.11c.

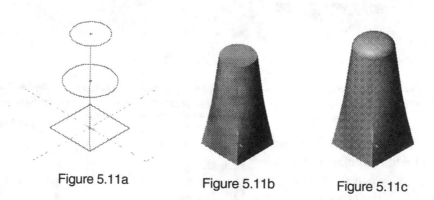

Figure 5.11a          Figure 5.11b          Figure 5.11c

### 5.5.2   Create the Loft Base Feature

Create the SWITCH. Click the **New** ◻ icon. Click **PartEnglishTemplate**. Click **OK**. Click the **Save** 🖫 icon. Enter part name. Enter **SWITCH**. Click **Save**.

Three *Sketch* planes are required to create the three *Cross Sections*. Create the three *Sketch* planes.

Create the first *Sketch* plane. The first *Sketch* plane is the default *Top* plane. Right-click the *Top* plane from the FeatureManager. Click **Show**. Display the *Isometric* view. Click the **Isometric** icon.

Create the second *Sketch* plane. Copy the *Top* plane. Hold the **Ctrl** key down. Click the **edge** of the *Top* plane. Drag the *Top* plane upward, Figure 5.12. Create the offset *Sketch* plane. Release the **mouse button**. Release the **Ctrl** key, Figure 5.13. Repaint the screen. Click the **Repaint** icon.

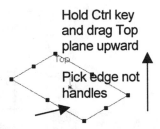

Figure 5.12

Note: Drag the *Top* plane by its edge or text name. Do not select the handles. The handles will resize the plane.

Create the third *Sketch* plane. Copy the *Top* plane. Hold the **Ctrl** key down. Click the **edge** of the *Top* plane. Drag the *Top* plane upward above *Plane1*. Release the **mouse button**. Release the **Ctrl** key. The default name for the third *Sketch* plane is *Plane2*, Figure 5.14.

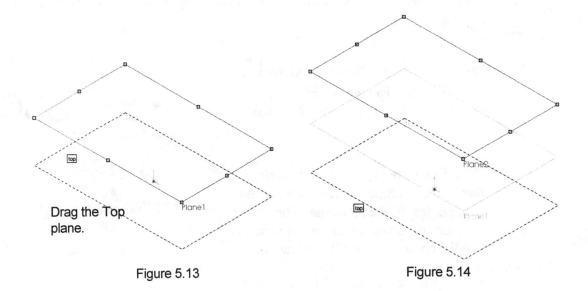

Figure 5.13                              Figure 5.14

Dimension the offset distance.  Double-click *Plane1*.  Enter **0.500**.

Double-click *Plane2*. Enter **1.000**.  Click **Rebuild**.

Display the planes from the *Front* view.  Click the **Front** ⬚ icon, Figure 5.15.

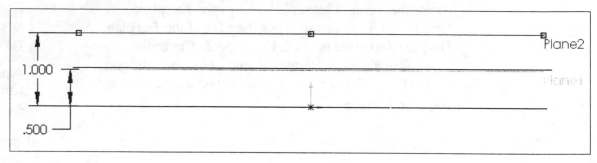

Figure 5.15

Display the planes in the *Isometric* view.  Click the **Isometric** ⬚ icon.

Note: Use the view icons to verify the location of the planes and cross sections.
Click the ***Front*** plane from the FeatureManager.  Click the ***Right*** plane.

Create *Sketch1* on the *Top* plane.  The profile of *Sketch1* is a square.

Select the *Sketch* plane.  Click the ***Top*** plane.  Display the *Top* view.  Click the
**Top** ⬚ icon.

Create the *Sketch*.  Click the **Sketch** 🖉
icon.  Click the **Rectangle** ⬚ icon.
Sketch a **Rectangle** centered about the
*Origin*, Figure 5.16.

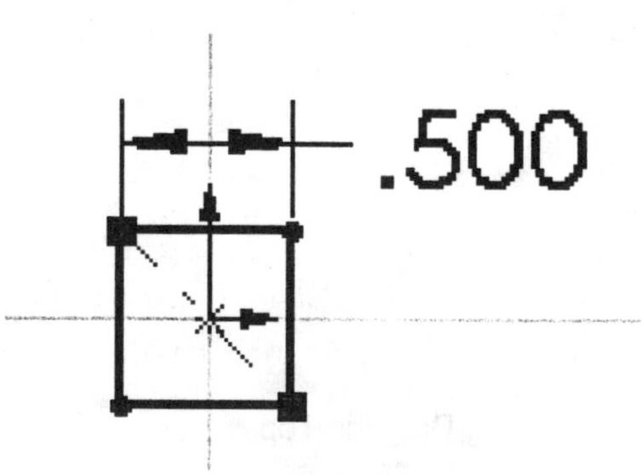

Click the **Centerline** ⦙ icon.  Create a
**diagonal** centerline between two opposite
corner points of the rectangle.  The
endpoints of the centerline are coincident
with the corner points of the rectangle.

Figure 5.16

Add geometric relations.  Click the **Add
Relations** ⊥ icon.

Add a **Midpoint** relation between the diagonal centerline and the *Origin*.  Click
**Apply**.

Add an **Equal** relation between the left vertical line and the top horizontal line. All four sides are equal. Click **Apply**. Click **Close**.

Add the dimensions. Click the **Dimension** [icon] icon. Dimension one line of the rectangle. Enter **0.500**.

Close *Sketch1*. Click the **Sketch** [icon] icon. Display the **Isometric** [icon] icon Figure 5.17.

Create *Sketch2* on *Plane1*, Figure 5.18a.

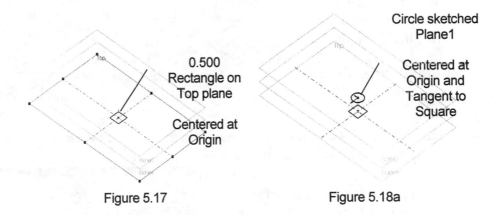

0.500
Rectangle on
Top plane

Centered at
Origin

Figure 5.17

Circle sketched
Plane1

Centered at
Origin and
Tangent to
Square

Figure 5.18a

Select the *Sketch* plane. Click *Plane1*.

Create the *Sketch*. Click the **Sketch** [icon] icon. Click the **Circle** [icon] icon. Create a **Circle** centered at the *Origin.*, Figure 5.18b.

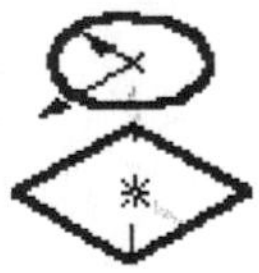

Display the *Top* view. Click the **Top** [icon] icon.

Figure 5.18b

Add geometric relations. Click the **Add Relations** [icon] icon. Click the **circumference** near the tangent. Click the *Sketch1* **line**. Click **Tangent**, Figure 5.19. Click **Apply**. Click **Close**.

Create a Tangent
relation between
the circle and line.

Close *Sketch2*. Click the **Sketch** [icon] icon.

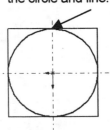

Figure 5.19

Create *Sketch3* on *Plane2,* Figure 5.20a. Select the *Sketch* plane. Click **Plane2.**

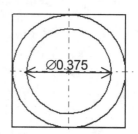

Ø0.375
Sketched on
Plane2

Centered
at Origin

Create the *Sketch*. Click the **Sketch** ✏ icon.
Click the **Circle** ⊕ icon. Create a **Circle** centered at the *Origin*, Figure 5.20b.

Figure 5.20a

Display the *Top* view.

Click the **Top** ⊞ icon.

Click the **Dimension** icon. Dimension the Circle. Enter **0.375**, Figure 5.20c.

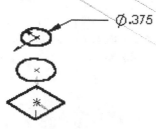

Ø.375

Ø0.375

Close *Sketch3*. Click the **Sketch** ✏ icon.

Figure 5.20b

Figure 5.20c

Hide the Planes. Click **View** from the Main menu. Click **Planes**. Fit the model to the Graphics window. Press the **f** key.

Create the *Loft*. Click the **Loft** 🔔 icon from the Feature toolbar. The Loft dialog box is displayed, Figure 5.21. Display the *Isometric* view. Click the **Isometric** 🧊 icon.

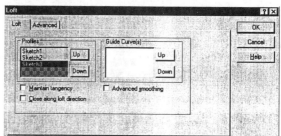

Figure 5.21

Select the Profiles. Click the **front corner** of *Sketch1*. Click **Sketch2**.

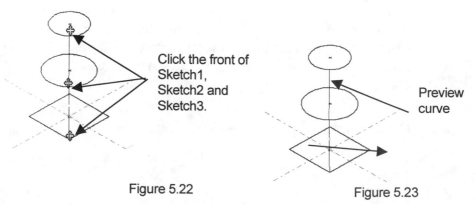

Click the front of
Sketch1,
Sketch2 and
Sketch3.

Preview
curve

Figure 5.22

Figure 5.23

Click **Sketch3**, Figure 5.22.

Note: Caution should be taken when selecting the profiles and location. The *Loft* feature is created by connecting the clicked locations of each profile. The system displays a preview curve as the profiles are selected, Figure 5.23.

The Up button and Down button in the Loft dialog box is used to rearrange the order of the profiles. The correct order for the profiles is:

- *Sketch1*

- *Sketch2*

- *Sketch3*

Figure 5.24

Note: If you select an incorrect location on the *Sketch*, Right-click in the Graphics window. Click Clear Selections. Select the profiles at the locations illustrated in Figure 5.22.

Display the *Base-Loft* feature, Figure 5.24. Click **OK**.

**Save** the SWITCH.

Create a *Dome* feature on the top face of the *Base-Loft* feature. Click the **top face** of the *Base-Loft* feature, Figure 5.25a.

Click the **Dome** icon from the Feature toolbar. The Dome dialog box is displayed. Enter **0.100**. Display the *Dome*, Figure 5.25b. Click **OK**.

Figure 5.25a          Figure 5.25b

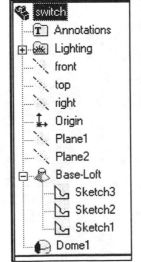

Figure 5.26

Expand the *Base-Loft* feature, Figure 5.26. The individual Sketches are displayed.

Modify the *Base-Loft* feature. Double-click on the **Base-Loft**, Figure 5.27. Double-click on the *Plane1* offset dimension, **0.500**. Enter **0.125**.

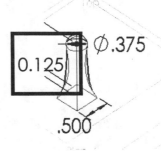

Figure 5.27

Click **Rebuild**, Figure 5.28.

**Save** the SWITCH.

Figure 5.28

## 5.6 LENSCAP

The LENSCAP is a plastic part used to hold the LENS to the HOUSING. An O-RING is positioned between the LENS and LENSCAP, Figure 5.29.

How should you sketch the LENSCAP? Sketch the LENSCAP on the *Front* plane.

What key dimensions from the LENS are required to create the LENSCAP?

The diameter and depth are the key dimensions to create the LENSCAP.

How do you determine the LENSCAP depth? Measure the depth of the LENS. The LENS is approximately 3" deep.

Position half of the LENS inside the LENSCAP. Position the other half within the HOUSING, Figure 5.30.

Figure 5.29

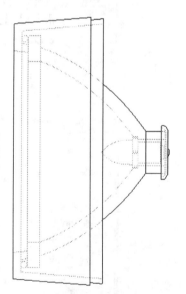

Figure 5.30

### 5.6.1    LENSCAP Overview

Create an *Extruded-Base* feature with a circular profile on the *Front* plane,
Figure 5.31a.  The LENSCAP is a plastic part.  For plastic parts us a Draft angle.

Add an *Extruded-Cut* feature.  The *Extruded-Cut* feature should be equal to the
diameter of the LENS *Revolved-Base* feature, Figure 5.31b.  Create a *Shell* feature.
Use the *Shell* feature for a constant wall thickness, Figure 5.31c.

Add a *Revolved-Cut* feature on the back face, Figure 5.31d.

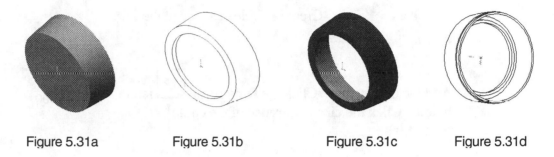

Figure 5.31a          Figure 5.31b          Figure 5.31c          Figure 5.31d

Create an *Extruded-Thin-Cut* feature.  The *Extruded-Thin-Cut* feature is used to
grip the outside of the LENSCAP, Figure 5.31e.  Create a *Pattern* feature.  The
*Pattern* feature is used to create multiple instances of the *Extruded-Cut* feature,
Figure 5.31f.

Create a thread using a sketched *Helical Curve* feature and a *Sweep* feature,
Figure 5.31g.

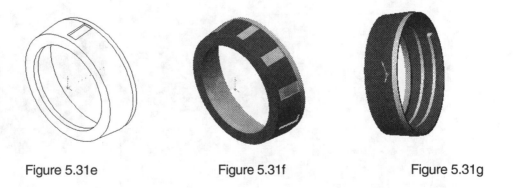

Figure 5.31e                    Figure 5.31f                    Figure 5.31g

### 5.6.2    Create the Extruded-Base Feature

Create the LENSCAP.  Click the **New** ⬜ icon.  Click **PartEnglishTemplate**.
Click **OK**.  Click the **Save** 💾 icon.  Enter the part name.  Enter **LENSCAP**.
Click **Save**.

Select the *Sketch* plane.  The *Front* plane is the
default *Sketch* plane.

Create the *Sketch*.  Click the **Sketch** icon.  Click
the **Circle** ⊕ icon.  Create a circle centered at the
*Origin* ↳, Figure 5.32.

Add the dimensions.  Click the **Dimension** icon
button.  Click the **circumference** of the circle.
Enter **4.900**.

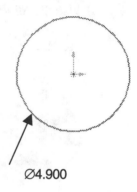

Ø4.900

Figure 5.32

Extrude the *Sketch*.  Click the **Base-Extrude** icon.  The Extrude Feature
dialog box is displayed, Figure 5.33.  Blind is the default Type option.

Enter **1.725** for Depth.  Click the **Reverse Direction** checkbox.  Click the **Draft
While Extruding** checkbox.  Enter **5** in the Angle text box.  Click the **Draft
Outward** check box.  Display the *Extruded-Base* feature.  Click **OK**, Figure 5.34.

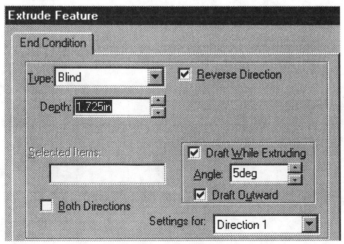

Figure 5.33                    Figure 5.34

**Save** the LENSCAP.

### 5.6.3   Create the Extruded-Cut Feature

Create an *Extruded-Cut* feature on the front face of
the *Base* feature.  The diameter of the *Cut* feature
equals the diameter of the *Base-Revolved* feature
of the LENS.

Select the *Sketch* plane.  Click the **front face**,
Figure 5.35.

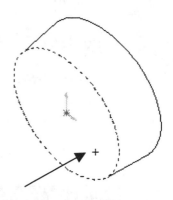

Create a *Sketch*.  Click the **Sketch** 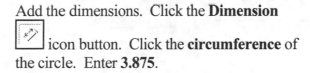 icon.  Click
the **Circle** icon.  Create a circle centered at the
*Origin*, Figure 5.36.

Figure 5.35

Add the dimensions.  Click the **Dimension**
icon button.  Click the **circumference** of
the circle.  Enter **3.875**.

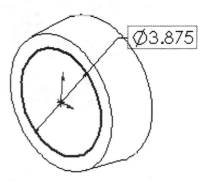

Extrude the *Sketch*.  Click the **Extruded-Cut**
icon.  Blind is the default Type option.
Enter **0.275** for Depth.  Click the **Draft While
Extruding** check box.  Enter **5** for Angle,
Figure 5.37.  Display the *Cut-Extrude* feature.
Click **OK**, Figure 5.38.

Figure 5.36

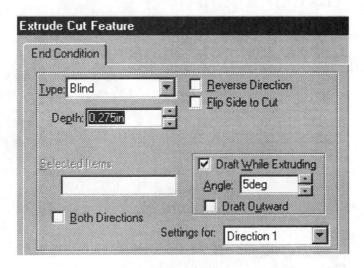

Figure 5.37

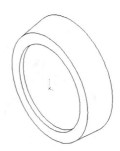

Figure 5.38

Rename *Cut-Extrude1* to *Front-Cut*.

**Save** the LENSCAP.

### 5.6.4   Create the Shell

The *Shell* feature removes face material from the solid LENSCAP.

Rotate the LENSCAP to view the part.  Click the **Rotate** [icon] icon.

Create the *Shell*.  Click the **Shell** [icon] icon from the Feature toolbar.  The Shell Feature dialog box is displayed, Figure 5.39.  Click the **front face** of the *Front-Cut* feature.  Click the **back face** of the *Extruded-Base* feature, Figure 5.40.

Enter **0.150** for Thickness.

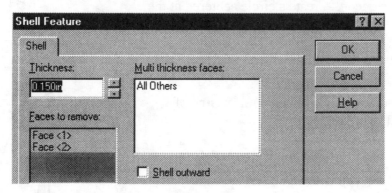

Figure 5.39

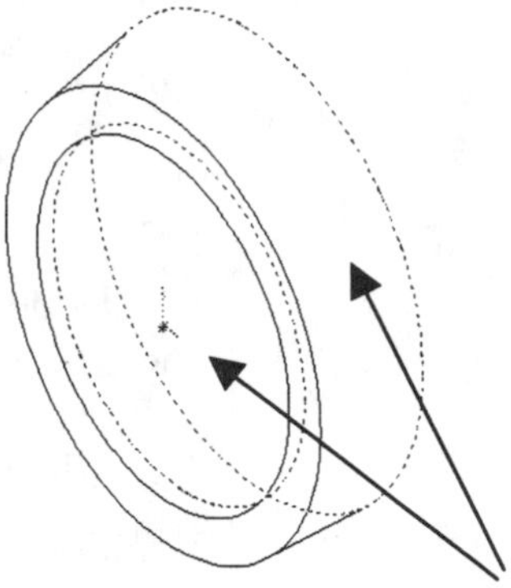

Figure 5.40

Display the *Shell*.  Click **OK**, Figure 5.41a.

Display the inside of the *Shell*.  Click the **Right** [icon] icon.  Click the **Hidden In Gray** [icon] icon, Figure 5.41b.

Figure 5.41a

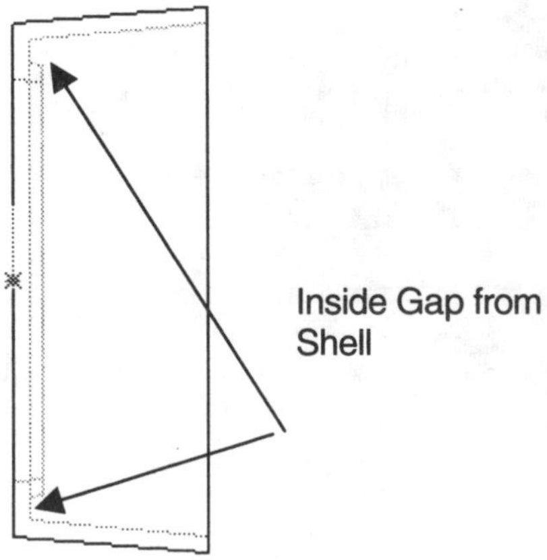

Inside Gap from Shell

The inside gap created by the *Shell* feature is used to seat the O-RING in the assembly.

**Save** the LENSCAP.

Figure 5.41b

### 5.6.5 Create the Revolved-Thin-Cut Feature

The *Revolved-Thin-Cut* feature removes material by rotating a sketched profile around a centerline.

Select the *Sketch* plane. Click the ***Right*** plane.

Create the *Sketch*. Click the **Sketch** icon.

Display the *Normal to* view. Click the **NormalTo** icon.

Sketch the centerline. Click the **Centerline** icon. Sketch a **horizontal centerline** collinear to the *Top* plane coincident to the *Origin*.

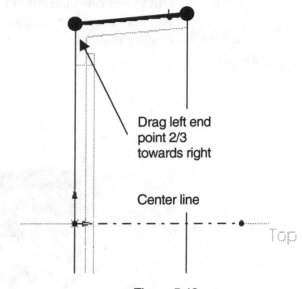

Drag left end point 2/3 towards right

Center line

Top

**Figure 5.42**

Sketch the profile. Right-click in the Graphics window. Click **Select**. Select the **top outside edge** of the *Extruded-Base* feature. Click the **Convert Entities** icon, Figure 5.42.

Create a short line. Drag the **left endpoint** 2/3 towards the right endpoint. Create the line. Release the **mouse button**, Figure 5.43.

Add dimensions. Click the **Dimension** icon. Click the **line**. Create an aligned dimension. Enter **0.250**.

Revolve the *Sketch*. Click the **Revolve Cut** icon from the Feature toolbar.

Note: Do not close the *Sketch*.

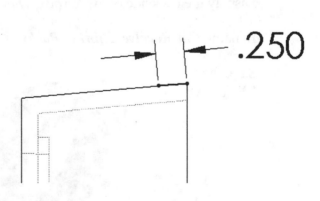

.250

**Figure 5.43**

The warning message states; "The sketch is currently open." Click **No**.

The Revolve Cut Thin Feature dialog box is displayed, Figure 5.44.

Create the feature on one side of the *Sketch*. Click the **Thin Feature tab**.
OneDirection is the default Type. Enter **0.050** for Wall Thickness. Click the
**Reverse** checkbox. Create the *Revolved-Thin* feature. Click **OK** from the Thin
Feature dialog box.

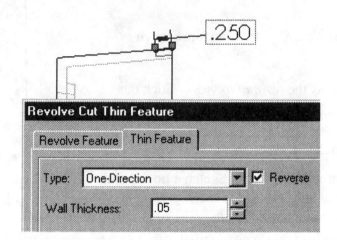

Figure 5.44

Display the *Revolve-Thin* feature, Figure 5.45a. Click **OK** from the Revolve
Feature dialog box.

Display the backside of the *Revolve-Thin* feature. **Rotate** the part, Figure 5.45b.

Rename *Cut-Revolve-Thin1* to *BackCut*.

**Save** the
LENSCAP.

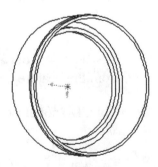

Figure 5.45a                                    Figure 5.45b

### 5.6.6  Create the Pattern

A *Pattern* creates one or more instances of a feature or group of features. A *Circular Pattern* requires a seed feature and an axis of revolution. The seed feature in this example is an *Extruded-Cut* feature created on a new *Surface Reference* plane. Create the following:

- A *Surface Reference* plane for the *Sketch* plane

- The seed feature as an *Extrude-Thin* feature

- The *Circular Pattern*

Display the planes in the *Isometric* view. Click the **Isometric** ⬡ icon.

Create a *Surface Reference* plane for the *Sketch* plane. Select the Plane Option. Click **Insert** from the Main menu. Click **Reference Geometry**. Click **Plane**. Click the **OnSurface** option, Figure 5.46. Click **Next**.

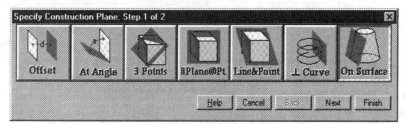

Figure 5.46

The On Surface dialog box is displayed, Figure 5.47a. Select the Surface. Click the **top outside conical face** of the *Extruded-Base* feature. Select a plane perpendicular to the surface. Click the *Right* plane from the FeatureManager, Figure 5.47b. Create the plane. Click **Finish**.

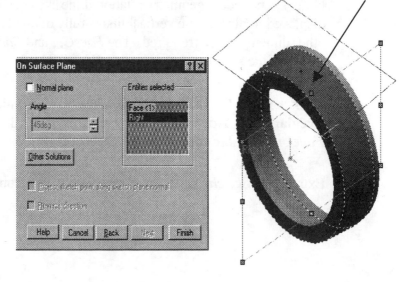

Figure 5.47a                                    Figure 5.47b

Rename *Plane1* to *SurfacePlane*.

**Save** the LENSCAP.

Display the *Right* plane. Right-click on the *Right* plane in the Feature Manager. Click **Show**.

Select the *Sketch* plane. Click the **SurfacePlane**.

Click the **NormalTo** ⚓ icon.

Create the *Sketch*. Click the **Sketch** 🖊 icon. Sketch a vertical line coincident to the *Right* plane.

Click the **Line** ╲ icon. Create a Line collinear with the *Right* plane. Create the first point. Click on the **intersection of the *Right* plane** and the *BackCut*.

Create the second point. Click on the **intersection of the *Right* plane** and the *FaceCut*, Figure 5.48.

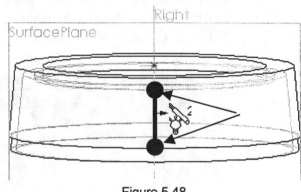

Figure 5.48

Note: Do not add a geometry relation if the sketched line and endpoints are displayed in black. The vertical line is fully defined and is displayed in black. No other dimensions are required. The *FaceCut* and *BackCut* features control the length of the seed feature.

Click the **Add Relations** ⊥ icon. Click the **first point**. Click the *BackCut* **circular edge**. Add a **Coincident** relation. Click the **second point**. Click the *FaceCut* **circular edge**. Add a **Coincident** relation. Click **Apply**. Click **Close**.

Extrude the *Sketch*. Click the *Extruded-Cut* ▣ icon.

Click the **Thin Feature** option from the Extrude as list box.  Click **Blind** from the Type drop down list.  Enter **0.100** for Depth.  Click the **Thin Feature** tab.  Enter **MidPlane** for Type.  Enter **0.500** for Wall Thickness, Figure 5.49a.

Display the *Extruded-Thin-Cut* feature, Figure 5.49b.  Click **OK** from the Thin Feature dialog box.

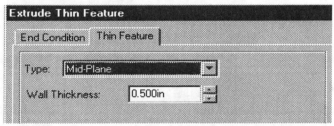

Figure 5.49a

Figure 5.49b

The *Cut-Extrude* is the seed feature for the *Pattern*.

Rename *Cut-Extrude* to *Seed-Cut*.

**Save** the LENSCAP.

Display the *Temporary Axis*.  Click **View** from the Main menu.  Click *Temporary Axis* from the View menu.

Create the *Pattern*.  Click the **Circular Pattern** ⬤ icon.  The Circular Pattern dialog box is displayed, Figure 5.50.  Create 10 cuts, equally spaced around the entire LENSCAP.  Enter **360** for Total angle.  Enter **10** for Total instances.  Click the **Equal spacing** check box.

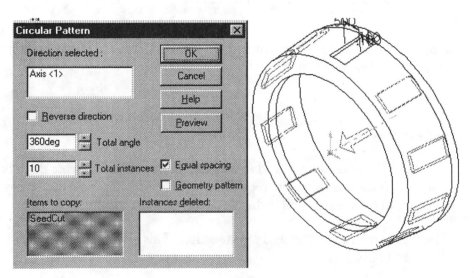

Figure 5.50

Display the *Circular Pattern*. Click **Preview**. Create the *Circular Pattern*. Click **OK**, Figure 5.51a.

Hide the Temporary axis. Click **View** from the Main menu. Click **Temporary axis**. Hide the Planes. Click **Planes**.

**Save** the LENSCAP.

Figure 5.51a

### 5.6.7  Suppress Features

A suppressed feature is a feature that is not displayed. It is useful to hide features for clarity. Suppressed features improve model Rebuild time.

Suppress the *SeedCut* feature. Right-click on the *SeedCut* feature in the FeatureManager. Click **Properties**. Click the **Suppress** checkbox, Figure 5.51b. Hide the feature. Click **OK**. The *SeedCut* and the *CirPattern* features are displayed in gray in the FeatureManager, Figure 5.51c.

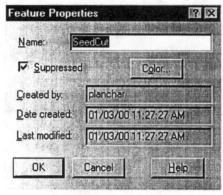
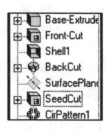

Figure 5.51b                                    Figure 5.51c

Note: The *Circular Pattern* feature is suppressed. The *CirPattern* feature is a child of the *SeedCut* feature.

### 5.6.8  Create the Sweep

The LENSCAP requires threads. Use the *Sweep* feature to create threads. The O-RING *Base-Sweep* feature required a circular path and a sketched cross section. The thread requires a spiral path. This path is called the *Threadpath*.

The thread requires a sketched cross section. This cross section is called the *Threadsection*.

Note: Coils and springs use helical curves.

There are numerous steps required to create a thread. The plastic thread on the LENSCAP requires a smooth lead in. The thread is not flush with the back face, Figure 5.52. Use an offset plane to start the thread.

Create a new offset *Sketch* plane, *ThreadPlane*. Thread cross sections are normally small and detailed. Create a thread. Below are the following steps:

- Create the thread path

- Create a large thread cross section for improve visibility

- Create the *Sweep* feature

- Reduce the size of the thread cross section

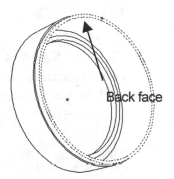

Figure 5.52

Create an *Offset Reference* plane for the *Sketch* plane. Display the back face of the LENSCAP. **Rotate** and **Zoom** the LENSCAP. Click the **narrow back face** of the *BackCut* feature.

Select the Plane Option. Click **Insert** from the Main menu. Click **Reference Geometry**. Click **Plane**. Click the **Offset** option. Click **Next**. The Offset Plane dialog box is displayed, Figure 5.53a. Enter **0.450** for Distance. Click the **Reverse** checkbox. Create the plane. Click **Finish**, Figure 5.53b.

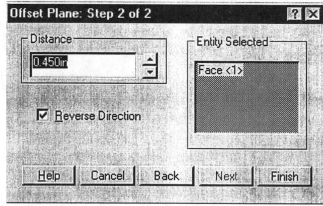

Figure 5.53a

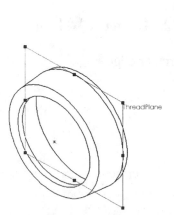

Figure 5.53b

Rename *Plane2* to *Threadplane*.

Display the *Isometric* view with the hidden lines removed. Click the **Isometric** icon. Click the **HiddenLinesRemoved** icon.

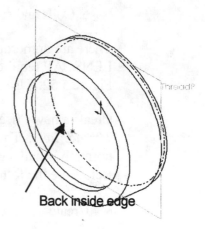

The current *Threadplane* is displayed in green.

Create the *Sketch*. Click the **Sketch** icon. Extract the edge to the *Threadplane*. Click the **back inside circular edge** of the *Shell*, Figure 5.54. Click the **Convert Entities** icon.

Back inside edge

Figure 5.54

The circular edge is displayed on the *Threadplane*, Figure 5.55. Verify the Sketch

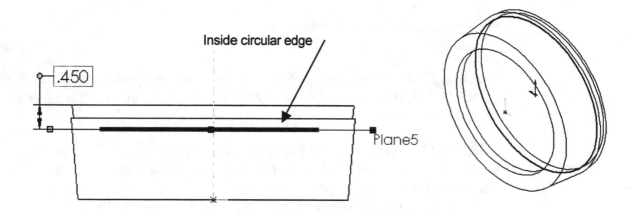

Inside circular edge

.450

Plane5

Figure 5.55

position. Click the **Top** icon. Click the **Isometric** icon.

Create the *Helix/Spiral* curve from the circular edge.

Create the *Helix/Spiral* curve path. Click **Insert** from the Main menu. Click **Curve**. Click *Helical/Spiral*. The Helix Curve dialog box is displayed, Figure 5.56. Enter **0.250** for Pitch. Enter **2.5** for Revolutions. Click the **Taper Helix** check box. Enter **5** for Angle. Enter **0** for Starting angle. Click the **Reverse direction** checkbox. Do not check the Taper outward check box.

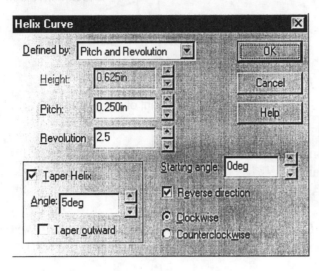

Figure 5.56

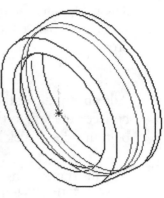

Figure 5.57

Display the *Helical/Spiral* path. Click **OK**, Figure 5.57.

Rename *Helix1* to *ThreadPath*.

Create the cross section.

Select the *Sketch* plane. Click the *Top* plane.

Create the *Sketch*. Click the **Sketch** icon. Display the *Top* view. Click the **Top** icon. Sketch the profile to the top right of the LENSCAP, Figure 5.58a.

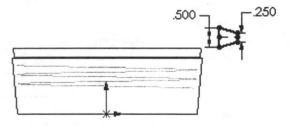

Figure 5.58a

Click the **Centerline** icon. Create a short horizontal **Centerline**, Figure 5.58b.

Sketch the profile. Click the **Line** icon.

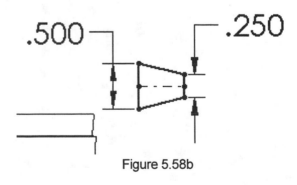

Figure 5.58b

Maintain the centerline as part of the *Sketch* with geometric relationships, Figure 5.59.  Click the **Add Relations** ⊥ icon.

Add a **Midpoint** relation between the left endpoint of the centerline and the sketched vertical line.

The Midpoint relation is required to pierce the thread cross section to the Threadpath.

Add an **Equal** relation between the left, top and bottom sides.  Click **Apply**.  Click **Close**.

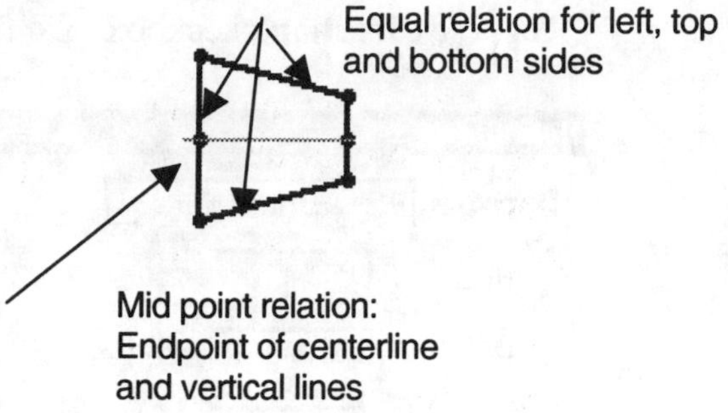

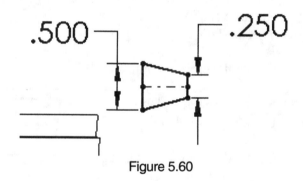

Equal relation for left, top and bottom sides

Mid point relation:
Endpoint of centerline and vertical lines

Figure 5.59

Add the dimensions.  Click the **Dimension** ✎ icon.  Click the **left vertical edge**.  Enter **0.500**.  Click the **right vertical edge**.  Enter **0.250**, Figure 5.60.

.500 ⌐                    ⌐ .250

Figure 5.60

Click the **Isometric** ⬡ icon

Click the **Add Relations** ⊥ icon.

Add a **Pierce** relation between the left midpoint of the cross section and the starting edge of the *Threadpath*, Figure 5.61.

The Pierce relation positions the midpoint of the cross section to the path. Display the *Sketch*. Click **Apply**. Click **Close**, Figure 5.62.

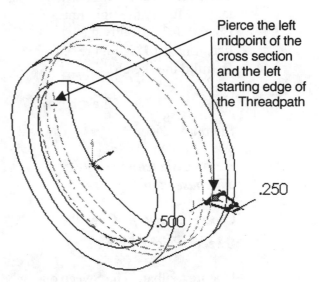

Pierce the left midpoint of the cross section and the left starting edge of the Threadpath

.250

.500

Figure 5.61

The cross section is inside the shell on the helix curve.
After the *Sweep* feature is applied, an internal thread is created.

Note: During the Pierce option, the selection location of the path determines the placement of the cross section. Example: If the helix is selected on the outer right side, an external thread is created. If the helix is selected on the inner left side, an internal thread is created.

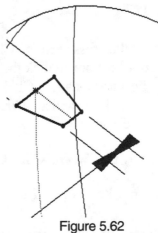

Figure 5.62

The thread cross section is too large. Modify the right line. Enter **0.063**. Modify the left line. Enter **0.125**, Figure 5.63. **Rebuild** the sketch.

Close the Sketch. Click the **Sketch** 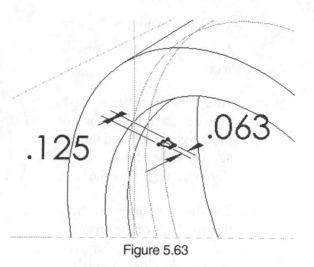 icon.

Rename *Sketch6* to *Threadsection*.

Create the *Sweep*. Click the **Sweep** icon from the

Feature toolbar. The Sweep dialog box is displayed, Figure 5.64.

Figure 5.63

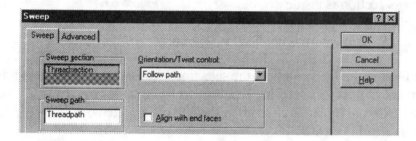

Figure 5.64

Select the cross section. Click the **Sweep section** text box. Click the *Threadsection* from the FeatureManager. Click the **Sweep path** text box. Click the helical *Threadpath* from the FeatureManager.

Display the *Sweep*. Click **OK**, Figure 5.65.

Rename *Boss-Sweep1* to *Thread.*

Expand the *Thread* feature. Click the **Plus Sign** icon in the FeatureManager, Figure 5.66. The *Thread* feature is composed of the *Threadsection* and *Threadpath*. Expand the ThreadPath. The Threadpath contains the circular *Sketch* and the definition of the Helical curve.

Note: If the *Threadsection* geometry intersects itself, the cross section is too big. Reduce the cross section size and create the *Sweep* feature again.

Figure 5.65

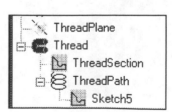

Figure 5.66

The LENSCAP is complete.

Restore the *CirPattern* feature.  Right-click on the ***CirPattern*** feature.  Remove the checkmark.  Click on the **Suppress** checkbox.

Figure 5.67

**Rebuild** the LENSCAP.  View the *Thread*.  **Rotate** the LENSCAP, Figure 5.67.

**Save** the LENSCAP.

## 5.7    HOUSING

The HOUSING provides storage and support for the BATTERY.  The SWITCH and LENSCAP connect to the HOUSING.

Do you remember the original customer requirements?

*   An inexpensive reliable flashlight

*   Available advertising space of 10 square inches

*   A light weight semi indestructible body

*   Self standing with handle

The HOUSING must meet the above customer requirements.

In a design situation, you may not have information on all of the required dimensions.  Where do you start?

The LENS, BATTERYPLATE and BATTERY are internal components to the HOUSING, Figure 5.68a.

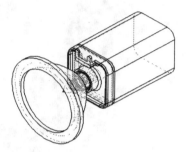

Figure 5.68a

The LENSCAP thread fastens to the HOUSING thread.

Create a circular *Extruded-Base* feature that fits inside the LENSCAP and contains the LENS.

In this exercise, create a loose fit between the BATTERY and the HOUSING.  Review the mating components and determine dimensions that require modification.

The LENS and the BATTERY are purchased parts.  You can not modify their size or shape!  How do you transition the HOUSING from a rectangular shape to a circular shape?  Answer: Create a *Loft* feature, Figure 5.68b.

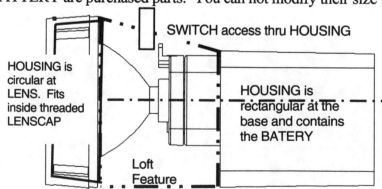

HOUSING is circular at LENS. Fits inside threaded LENSCAP

SWITCH access thru HOUSING

HOUSING is rectangular at the base and contains the BATERY

Loft Feature

Figure 5.68b

The HOUSING supports the threaded LENSCAP and LENS. The SWITCH fits through an opening in the HOUSING.

Internal *Ribs* are added to the HOUSING for structural integrity.

## 5.8   HOUSING Overview

The HOUSING is composed of the following features:

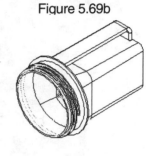

Figure 5.69a                    Figure 5.69b

- *Extruded-Base* feature, Figure 5.69a

- *Loft* feature, Figure 5.69b

- *Boss-Extrude* feature and *Shell* feature, Figure 5.69c

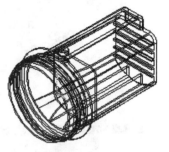

Figure 5.69c                    Figure 5.69d

- *Handle Sweep and Thread-Sweep* features, Figure 5.69d

- *Rib* feature and *Linear Pattern*, Figure 5.69e

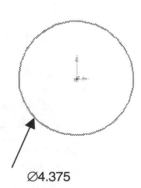

### 5.8.1   Create the Extruded-Base Feature

Figure 5.69e

Create the HOUSING. Click the **New** ⬜ icon. Click **PartEnglishTemplate**. Click **OK**. Click the **Save** 💾 icon. Enter part name. **Enter** HOUSING. Click **Save**.

Select the *Sketch* plane. The *Front* plane is the default *Sketch* plane.

Create the *Sketch*. Click the **Sketch** ✏ icon. Click the **Circle** ⊕ icon. Create a circle centered at the *Origin* ⌐, Figure 5.70a.

Ø4.375

Figure 5.70a

Add dimensions.  Click the **Dimension** 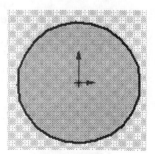 icon button.  Dimension the diameter of the circle.  Enter **4.375**.

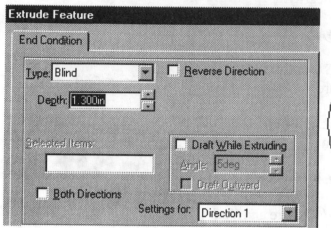

Figure 5.71a                                              Figure 5.71b

Extrude the *Sketch*.  Click the **Extruded-Base** icon.  The Extrude Feature dialog box is displayed, Figure 5.71a.  Blind is the default Type option.  Enter **1.300** for Depth.  Display the *Base-Extrude* feature.  Click **OK**, Figure 5.71b.

**Save** the HOUSING.

### 5.8.2  Create the Loft Feature

The *Loft* feature is composed of two profiles. Create the first profile from the back face of the *Base-Extrude* feature.

Create the second profile on an *Offset* plane. Close the *Sketch*.  Create the *Loft*.

Create the first profile.

Figure 5.72a

Select the *Sketch* plane.  Click the **back face** of the *Base-Extrude* feature, Figure 5.72a.

Create *Sketch2*.  Click the **Sketch** icon.  Extract the entire face.  Click the **Convert Entities** icon, Figure 5.72b.

**Close** *Sketch2*.

Figure 5.72b

Rename *Sketch2* to *SketchCircle*.

Create the second profile.

Create the *Sketch* plane. Click the **back face**. Click **Insert** from the Main menu. Click **Reference Geometry, Plane**. Create an **Offset** plane. Enter **1.300**, Figure 5.73a. Verify the plane position.

Click the **Top** 📐 icon, Figure 5.73b. Click **Finished**.

Rename *Plane4* to *batteryloftplane*.

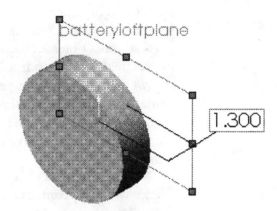

Figure 5.73a

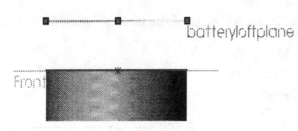

Figure 5.73b

Create the *Sketch*. Click the **Sketch** 🖊 icon. Click the **NormalTo** ⚓ icon. Extract the circular edge. Click the **circumference of the circle**. Click the **Convert Entities** 🗗 icon. Click the **Centerline** ┃ icon. Sketch a **centerline** collinear with the *Right* plane coincident to the *Origin*.

Click the **Mirror** 🔼 icon. Click the **Line** ⬟ icon. Sketch a **horizontal** line. Click the **Tangent Arc** 🗗 icon. Sketch a **90 degree arc**. Click the **Line** ⬟ icon. Sketch a **vertical** line. The endpoint of the vertical line is coincident with the edge of the circle, Figure 5.74a. Turn off the Mirror. Click the **Mirror** 🔼 icon.

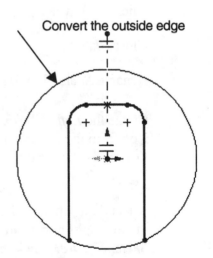

Convert the outside edge

Figure 5.74a

Add dimensions. Click the **Dimension** ✐ icon. Create the **horizontal** dimension. Click the **left vertical line**. Click the **right vertical line**. Enter **3.100**. Create the **vertical** dimension from the *Origin*. Enter **1.600**. Create the **radial** dimension. Enter **0.500**, Figure 5.74b. The FLASHLIGHT components must remain aligned to a common centerline.

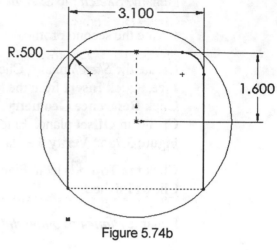

Figure 5.74b

Trim unwanted geometry. Click the **Trim** ⌖ icon. Click the far **right edge of the circle**. Click the far **left edge of the circle**, Figure 5.74c.

Remove sharp edges. Add fillets at the lower corners. Click the **Select** ▷ icon. Click the **lower left point**. Click the **2D Fillet** icon. Enter **0.500**. Click **Apply**. Click **the lower right**. Click **Close**.

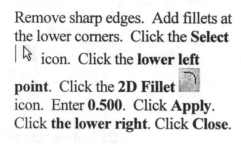

Close *Sketch3*.

Rename *Sketch3* to *SketchSquare*.

Create the *Loft*. Click the **Loft** 🖌 icon from the Feature menu. The Loft dialog box is displayed, Figure 5.75a.

Figure 5.74c

Select the Profiles. Click the **top right corner of** *SketchSquare*. Click the **upper right side of the** *SketchCircle*, Figure 5.75b.

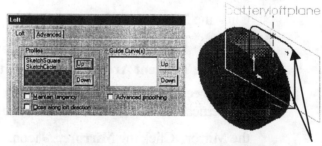

Figure 5.75a　　　　　Figure 5.75b

Display the *Loft* feature.
Click **OK**, Figure 5.75c.

The *Boss-Loft1* feature is
composed of *SketchSquare*
and *SketchCircle*,
Figure 5.75d.

**Save** the HOUSING.

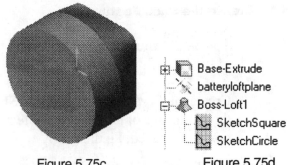

Figure 5.75c                 Figure 5.75d

### 5.8.3   Create the First Extruded-Boss Feature

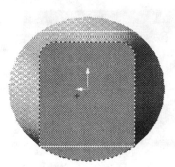

Create the first *Extruded-Boss* feature from the square
face of the *Loft*. How do you estimate the depth of this
feature?

Answer: The *Extruded-Base* feature of the BATTERY
is 4.1". *Ribs* are required to support the BATTERY.
Design for *Rib* construction. Use a 4.4" depth as the
first estimate.

Note: Adjust the estimated depth dimension later if
required in the FLASHLIGHT assembly.

Figure 5.76

Select the *Sketch* plane. Click the **back face** of the *Loft*, Figure 5.76.

Create the *Sketch*. Click the **Sketch** [icon] icon. Extract the entire face. Click the
**Convert Entities** [icon] icon.

Extrude the *Sketch*. Click the **Extruded-Boss/Base** [icon] icon. Enter **4.400** for
Depth. Check **Draft While Extruding** check box. Enter **1** for Draft Angle,
Figure 5.77a. Display the *Boss* feature. Click **OK**, Figure 5.77b.

Rename *Boss-Extrude* to ***Boss-Battery***. **Save** the HOUSING.

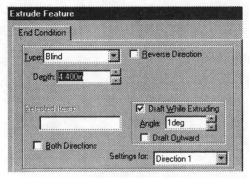

Figure 5.77a

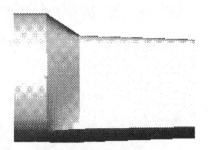

Figure 5.77b

### 5.8.4   Create the Shell Feature

The *Shell* feature removes material.  Use the *Shell* feature to remove the front face of the HOUSING.

Create the *Shell* feature.  Click the **Shell** ⬛ icon from the Feature toolbar.  Click the **front face** of the *Boss-Insert* feature, Figure 5.81a.  The Shell Feature dialog box is displayed, Figure 5.81b.

Enter **0.100** for Thickness.

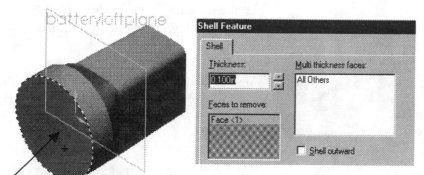

Figure 5.81a

Figure 5.81b

Display the *Shell*.  Click **OK**, Figure 5.81c.

Note: First create the *Shell* feature.  Then create the S*weep* feature for the handle.  The handle is solid.

**Save** the HOUSING.

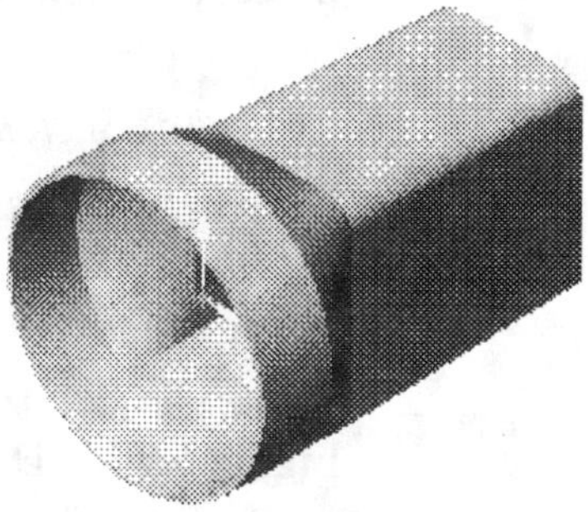

### 5.8.5   Create the Second Extruded-Boss Feature

The second *Extruded-Boss* feature creates a solid circular ring on the back circular face of the *Extruded-Base* feature.  The solid ring is a cosmetic stop for the LENSCAP and provides rigidity at the transition of the HOUSING.

Figure 5.81c

Select the *Sketch* plane.  Click the **Front** plane.

Create the *Sketch*, Figure 5.82a. Click the **Sketch** ▨ icon. Create the inside circle. Click the **front inside circular edge** of the *Shell*. Click the **Convert Entities** ▱ icon.

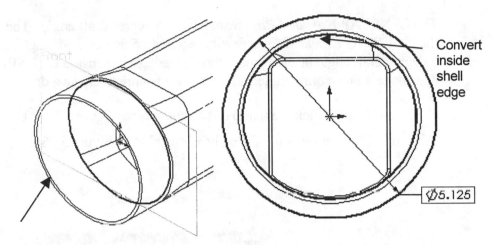

Figure 5.82a

Create the outside circle. Click the **Circle** ⊕ icon. Create a **circle** centered at the *Origin*. Add dimensions. Click the **Dimension** ◇ icon. Click the outside circle. Enter **5.125**.

Extrude the *Sketch*, Figure 5.82b. Click the **Extruded Boss/Base** ▣ icon. Enter **0.100** for Depth. Display the *Boss* feature. Click **OK**, Figure 5.82c.

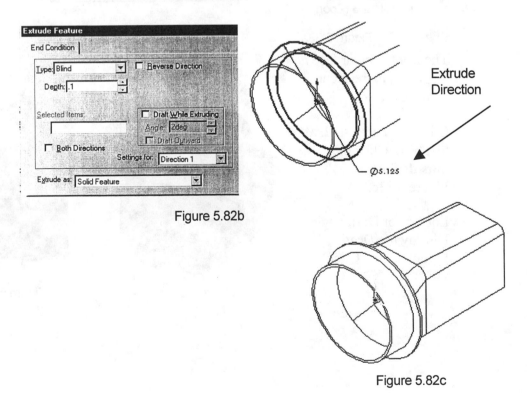

Figure 5.82b

Figure 5.82c

Rename *Boss-Extrude* to *Boss-Stop*.

**Save** the HOUSING.

### 5.8.6   Create the Draft Feature

The LENSCAP *Extruded-Base* feature has a 5 degree draft angle. The outside of
the HOUSING *Extruded-Base* feature requires a 5 degree draft angle. The
5 degree draft angle insures proper thread mating. The inside HOUSING wall
does not contain a draft angle. Use the *Draft* feature to create a draft angle.

Create the *Draft*. Click the thin **front circular face**, Figure 5.83a. Click the **Draft**
icon. The Draft Feature dialog box is displayed, Figure 5.83b.

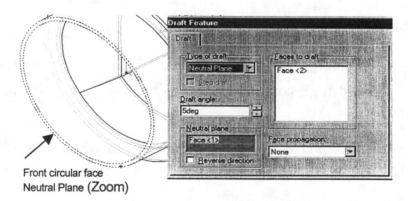

Front circular face
Neutral Plane (Zoom)

Figure 5.83a                              Figure 5.83b

The front circular face is
displayed in the Neutral
plane text box.

Click the **Faces to draft**
text box. Click the
**outside face**,
Figure 5.83c.

Enter **5** for Draft angle.
Display the *Draft*,
Figure 5.83d. Click
**OK**.

Outside face to draft

Figure 5.83c                              Figure 5.83d

Display the draft angle and the straight interior. Click the **Right** view. Click **Hidden Lines Removed**, Figure 5.83e.

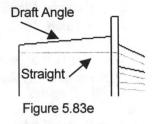

Figure 5.83e

**Save** the HOUSING.

The threads for the HOUSING are created on the outside drafted face.

### 5.8.7   Create the Thread with the Sweep Feature

The HOUSING requires a thread. Create a thread with the *Sweep* feature. The thread requires two sketches:

1. *Threadpath*

2. *Threadsection*

Create the *Threadpath* sketch. Create a *Helix* curve offset from the front face.

Create an *Offset Reference* plane. Click **Insert** from the Main menu. Click **Reference Geometry**. Click **Plane**. Click **Offset**. Click **Next**. Click the **front circular face** of the HOUSING, Figure 5.84a. Click the **Reverse** check box. Enter **0.125** for Distance. Click **OK**.

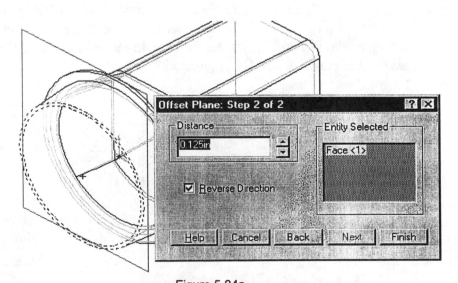

Figure 5.84a

Rename *Plane* to
*Threadplane*.

Verify the *ThreadPlane*
position. Display the Top

view. Click the **Top** 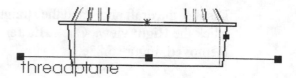 icon,
Figure 5.84b.

Figure 5.84b

Select the *Sketch* plane.
Click the *Threadplane*.

Create the *Sketch*. Click

the **Sketch** icon.
Select the **front outside
circular edge**,
Figure 5.85a.

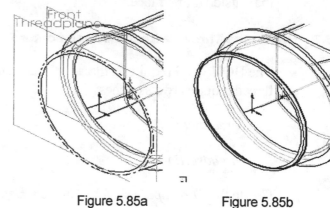

Click the **Convert**                 Figure 5.85a                Figure 5.85b

**Entities** icon. The
circular edge is displayed on the *Threadplane,* Figure 5.85b.

Create the *Helix/Spiral* curve from the circular edge. Click **Insert** from the Main
menu. Click **Curve**. Click *Helix/Spiral*. The Helix Curve dialog box is
displayed, Figure 5.86.

Enter **0.250** for Pitch. Enter **2.5** for Revolutions. Click the **Taper Helix** check
box. Enter **5** for Angle. Click the **Taper Outward** check box. Enter **180** in the
Starting angle spin box. Click the **Reverse direction** checkbox. Display the
*Helix/Spiral* curve. Click **OK**, Figure 5.87.

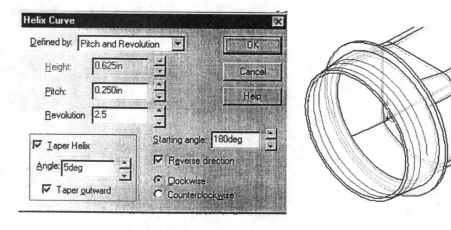

Figure 5.86                                    Figure 5.87

Rename *Helix1* to *Threadpath*. **Save** the HOUSING.

You have just created the *Threadsection* for the LENSCAP.

Create the *Threadsection* sketch for the HOUSING. Copy the LENSCAP created sketched cross section.

**LENSCAP**

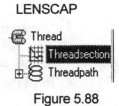

**Figure 5.88**

**Open** the LENSCAP. **Expand** the *Thread-Sweep* feature from the FeatureManager. Click the ***Threadsection*** sketch, Figure 5.88. Click **Edit** from the Main menu. Click **Copy**. **Close** the LENSCAP.

Click the ***Top*** plane in the HOUSING FeatureManager. Click **Edit** from the Main menu. Click **Paste**. The *Threadsection* is displayed on the *Top* plane, Figure 5.89a. The new *Sketch7* name is added to the FeatureManager.

Rename ***Sketch7*** to ***Threadsection***.

Pierce *Threadsection* to *Threadpath*. Right-click on ***Threadsection***. Click **Edit Sketch**.

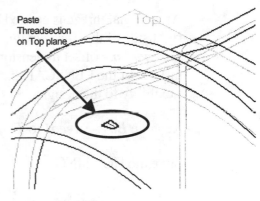

**Figure 5.89a**

Add a Pierce relation. Click the **Add Relations** ⃫ icon.

Enlarge the *Sketch* for improve visibility. Click the ***ThreadSection***. Click the **Zoom to Selection** 🔍 icon, Figure 5.89b. Click the **Midpoint** of the *Threadsection*.

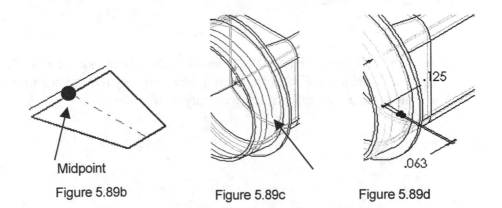

Midpoint

Figure 5.89b                    Figure 5.89c                    Figure 5.89d

Display the *Threadpath*. Click the **Zoom to Area** 🔍 icon. Click the **right back edge of the *Threadpath***, Figure 5.89c. The Pierce relation positions the center of the cross section on the path, Figure 5.89d. Click **Pierce**. Click **Apply**. Click **Close**. Caution: Do not click the front edge of the *Thread path*. The *Thread* is created out of the LENSCAP.

Close the *Sketch*.  Click the **Sketch** icon.

Sweep the *Sketch*.  Click the **Sweep** icon from the Feature toolbar.  Select the cross section.  Click the **Sweep section** text box.  Click *Threadsection* from the FeatureManager.  Click the **Sweep path** text box.  Click the helical *Threadpath* from the FeatureManager.

Display the *Sweep*.  Click **OK**, Figure 5.90.

An external thread is created by piercing the midpoint of the *Threadsection* to the right side of the *Threadpath*.  The *Threadplane* allows for a smooth lead.  The *Threadplane* offset dimension is modified in the assembly by adjusting the mating threads of the LENSCAP and HOUSING.  Creating a *Threadplane* provides flexibility to the design.

Rename *Boss-Sweep1* to *Thread*.

**Save** the HOUSING.

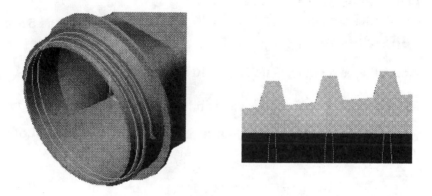

Figure 5.90

Note: Conserve time.  Use the Right-mouse button to invoke the Zoom and Pan commands.  Selection Filters are helpful, however they require one or more clicks.  View the mouse pointer feedback for entity confirmation.

### 5.8.8   Create the Handle with the Sweep Feature

Create the handle with the *Sweep* feature. The *Sweep* feature consists of a sketched path and cross section. Sketch the path on the *Right* plane. The sketch uses edges from existing features. Sketch the cross section on the back circular face of the *Boss-Stop* feature.

Create the *Sweep* path.

Select the *Sketch* plane.
Select the **Right** plane.

Create the *Sketch*. Click the **Sketch** icon. Click the **top small horizontal edge**. Click the **Convert Entities** icon. Click the **top right vertical small edge**. Click the **Convert Entities** icon, Figure 5.91.

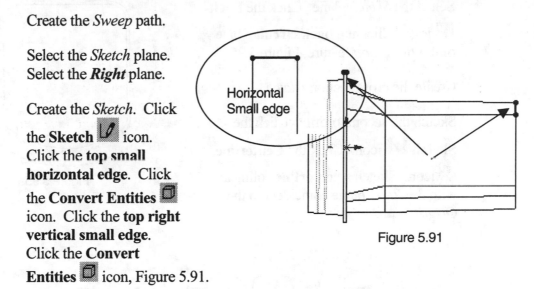

Figure 5.91

Drag the **left point** on the horizontal line toward the back edge, Figure 5.92. Drag the **bottom point** on the vertical line upward to complete the intersection, Figure 5.93.

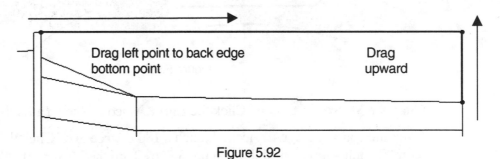

Figure 5.92

Create a 2D Fillet. Click the **upper right intersection**. Click the **2D Fillet** icon. Enter **0.500** for Radius, Figure 5.94. Click **Close**.

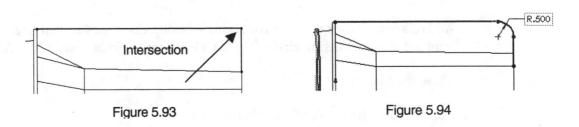

Figure 5.93                                     Figure 5.94

Close the *Sketch*. Click the **Sketch** icon.

Rename *Sketch8* to *HandlePath*.

Create the *Sweep* cross section.

Select the *Sketch* plane. Click the **Back** icon. Click the **back circular face** of the *Boss-Stop* feature, Figure 5.95a.

Create the cross section *Sketch*.

Sketch the second profile. Click the **Sketch** icon. Click the **Centerline** icon. Sketch a **centerline** collinear with the *Right* plane coincident to the *Origin*.

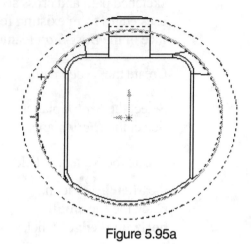

Figure 5.95a

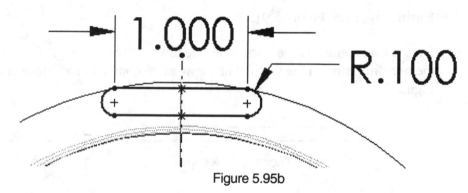

Figure 5.95b

Click the **Mirror** icon. Click the **Line** icon. Sketch a **horizontal** line. Click the **Tangent Arc** icon. Sketch a **180 degree arc**. Click the **Line** icon. Sketch a **horizontal** line, Figure 5.95b. Turn the Mirror off. Click the **Mirror** icon.

Add dimensions. Click the **Dimension** icon. Create the **horizontal** and **radial dimensions.**

Add relations. Click the **Add Relations** icon. Click the **tangent point** of the right arc and the **outside circle**. Click **Coincident**. Click **Close**. Click **Apply**.

**Close** the Sketch.

Rename *Sketch9* to *HandleSection*.

Sweep the *Sketch*. Click the **Sweep** icon from the Feature toolbar. The Sweep dialog box is displayed, Figure 5.95c.

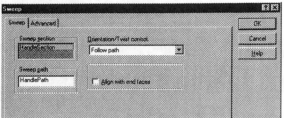

Figure 5.95c

Select the *Sweep* section. Click the **Sweep section** text box. Click the ***HandleSection*** from the FeatureManager. Click the **Sweep path** text box. Click the ***HandlePath*** from the FeatureManager. Display the *Sweep*. Click **OK**, Figure 5.95d.

Rename *Sweep2* to *Handle*.

**Save** the HOUSING.

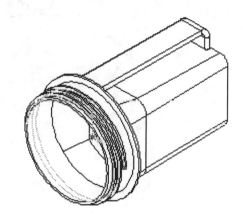

### 5.8.9    Create the Extruded-Cut Feature

Figure 5.95d

Insert the SWITCH into the HANDLE of the HOUSING.

Create an *Extruded-Cut* in the HANDLE for the SWITCH.

Select the *Sketch* plane. Click the **top face** of the *Handle*, Figure 5.95e.

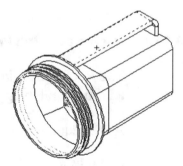

Figure 5.95e

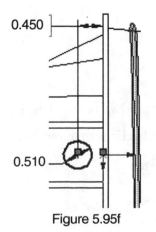

0.450

0.510

Figure 5.95f

Create the *Sketch*. Click the **Sketch** icon, Figure 5.95f. Create a circle. Click the **Circle** icon. Center the **circle** collinear to the ***Right*** plane. Add dimensions. Click the **Dimension** icon. Enter **0.510** for diameter. Enter **0.450** for the distance from the *Front* plane.

Extrude the *Sketch*. Click the **Extruded-Cut** 🔲 icon. Click the **UpTo Surface** option from the Type list box. Click the top inside face of the *Shell*, Figure 5.95g. Display the *Extruded-Cut*, Figure 5.95h. Click **OK**.

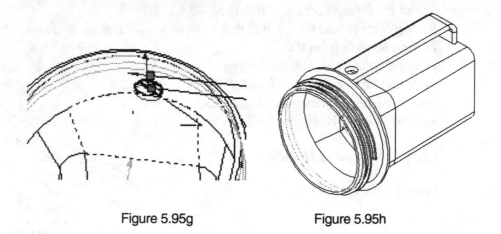

Figure 5.95g                              Figure 5.95h

Rename *Extrude-Cut1* to *SwitchHole*.

**Save** the HOUSING.

### 5.8.10 Create the First Rib Feature

The *Rib* feature adds material between contours of existing geometry. Use *Ribs* to add structural integrity to a part.

A *Rib* requires:

- A *Sketch*

- Thickness

- Extrusion direction

Create the *Rib*. Select the *Sketch* plane. Click the **Top** plane. Display the Top view. Click the **Top** 🔲 icon. Display all hidden lines to avoid unwanted relationships. Click the **Hidden In Gray** 🔲 icon.

Create the *Sketch*. Click the **Sketch** icon. Click the **Line** icon. The endpoints of the sketch are collinear with the inside wall of the *Shell* feature, Figure 5.95i.

Add a linear dimension. Click the **Dimension** icon. Click the **line**. Click the **inner back edge**. Enter **0.175**.

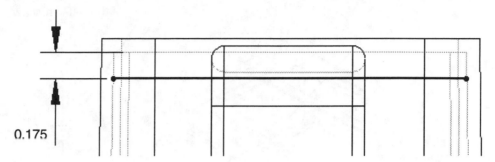

0.175

Figure 5.95i

Create the *Rib*. Click the **Rib** icon from the Feature toolbar. The Rib Property box is displayed, Figure 5.95j.

Create the *Rib* on both sides of the *Sketch* plane. Click the **Mid plane** text. Enter **0.075** for Thickness.

Click the **Flip material side** check box. The Rib direction arrow points to the back, Figure 5.95k.

Click the **Enable draft** check box. Click the **Draft outward** check box.

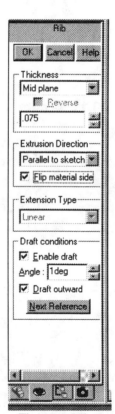

Figure 5.95j

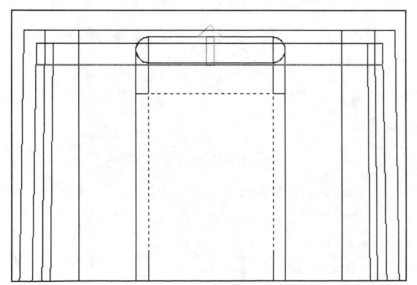

Figure 5.95k

Display the *Rib*. Click the **Isometric** 🔲 icon. Click **OK**, Figure 5.96. Click the **Right** 🔲 icon, Figure 5.97.

Existing geometry defines the *Rib* boundaries. The *Rib* does not penetrate through the wall.

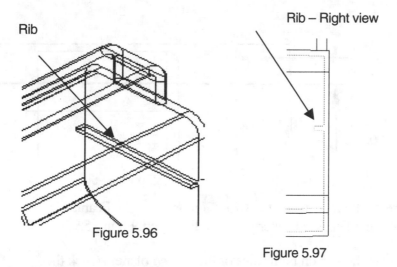

Rib

Figure 5.96

Rib – Right view

Figure 5.97

### 5.8.11  Create a Linear Pattern of Ribs

The HOUSING requires multiple *Ribs* to support the BATTERY. A *Linear Pattern* creates multiple Instances of a feature along a straight line.

Create a *Linear Pattern* of the *Rib* feature. Click the **Linear Pattern** 🔲 icon. The Linear Pattern dialog box is displayed, Figure 5.98.

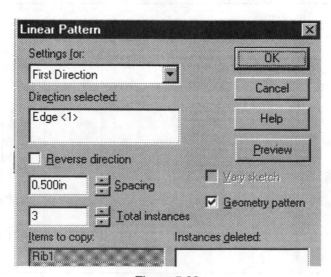

Figure 5.98

Click *Rib1*. The *Linear Pattern* requires a direction. Click the **Direction selected** text box. Click the **back vertical edge**, Figure 5.99.

Instances are created inside the HOUSING. Enter **0.500** for Spacing. Enter **3** for Total instances. *Rib1* is displayed in the Items to copy text box. Display the *Linear Pattern*. Click **Preview**. Click **OK**, Figure 5.100.

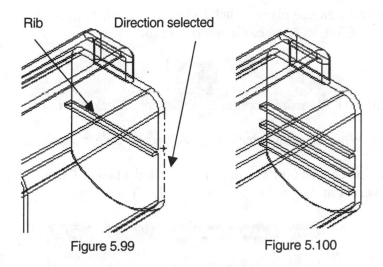

Figure 5.99                                                        Figure 5.100

Create additional *Ribs* for support.

Click **Insert** from the Main menu. Click **Pattern/Mirror**. Click the **Mirror** Feature.

The Mirror Pattern Feature dialog box is displayed, Figure 5.101. Click the **Mirror plane** text box. Click the *Top* plane.

The *LPattern1* feature is displayed in Feature to mirror text box. Click the **Geometry pattern** check box.

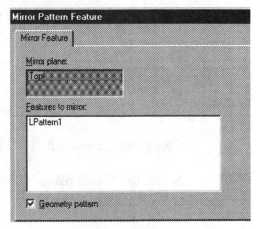

Figure 5.101

Display the *Mirror Pattern* feature, Figure 5.102. Click **OK**.

**Save** the HOUSING.

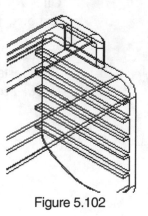

Figure 5.102

### 5.8.12  Create the Second Rib Feature

The Second *Rib* feature supports and centers the battery.

Create the *Sketch* plane. Click **Insert** from the Main menu. Click **Reference Geometry, Plane**.

Click the || **Plane@Pt** option   icon. Click **Next**.

Note: When selecting small edges and points do not use the Shaded view.

The Point-Plane option requires a point and a plane. Click the ***Right*** plane. Click the **vertex** at the back right of the handle, Figure 5.103.

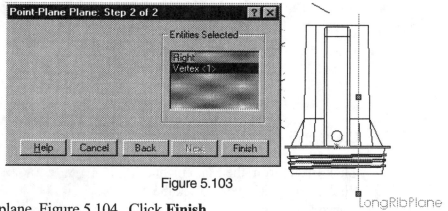

Figure 5.103

Figure 5.104

Display the plane, Figure 5.104. Click **Finish**.

Rename ***Plane1*** to ***LongRibPlane***.

Create the second *Rib*.

Select the *Sketch* plane. Click the ***LongRibPlane***. Display the *Right* view. Click the **Right**  icon.

Create the *Sketch*. Click the **Sketch** [icon] icon. Click the **Line** [icon] icon. Sketch a **horizontal line**. The right endpoint of the line is collinear with the inside back wall of the *Shell* feature, Figure 5.105.

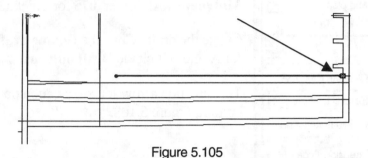

Figure 5.105

Note: When sketch and reference geometry become complex, create dimensions by selecting reference planes in the FeatureManager. For design flexibility, dimension the *Rib* from the *Origin* rather than from the bottom HOUSING surface. Do not sketch the *Rib* line collinear to the horizontal lines of the HOUSING. Click the Part [icon] icon to select *Planes* from the FeatureManager.

Click the **Dimension** [icon] icon. Click the **horizontal line**. Click the *Origin*. Click the location for the vertical linear dimension **text**, Figure 5.106. Enter **1.300**.

Click the **Tangent Arc** [icon] icon. Click the **left end** of the horizontal line. Click the **intersection of the inside wall**. Click the *Boss-Stop* feature.

Click **Add Relations** [icon] icon. Click the **end point** of the arc. Click the intersection of the **horizontal inside wall**. Click the **vertical *Boss-Stop* edge**.

Add a coincident relation between the left end point of the line and the *batteryloftplane*. Click the **left end point**. Click the *batteryloftplane* from the FeatureManager. Click **Apply**. Click **Close**.

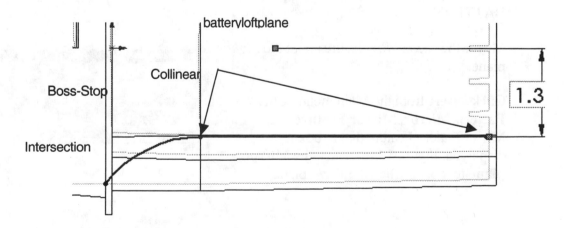

Figure 5.106

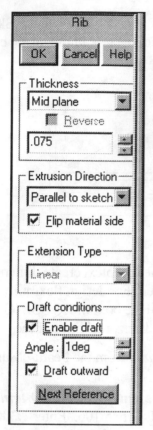

Figure 5.107

Create the *Rib*. Click the **Rib** 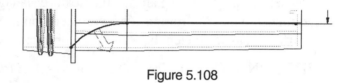 icon from the Feature toolbar. The Rib Property Manager is displayed, Figure 5.107.

Create the *Rib* on both sides of the *Sketch* plane. Click the **Mid plane** text. Enter **.075** for Thickness.

Create the draft. Click the **Enable draft** check box. Enter **1** for Angle. Click the **Draft outward** check box.

The direction arrow for *Rib* creation points towards the bottom, Figure 5.108.

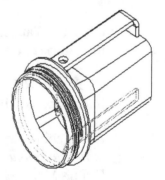

Figure 5.108

Display the *Rib*. Click the **Isometric** icon. Click **OK**, Figure 5.109.

### 5.8.13 Mirror the Second Rib

An additional *Rib* is required to support the BATTERY.

Figure 5.109

Mirror the second *Rib* feature about the *Right* plane.

Click **Insert** from the Main menu. Click **Pattern/Mirror**, **Mirror Feature**. The Mirror Pattern Feature dialog box is displayed, Figure 5.110. Click the **Mirror plane** text box. Click the *Right* plane.

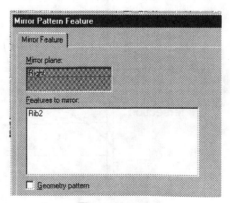

Figure 5.110

The *Rib2* feature is displayed in the
Feature to mirror text box.

Display the *Mirror Pattern* feature,
Figure 5.111a.  Click **OK**.

Note: Adjust *Ribs* in the assembly
to improve fit.

**Save** the HOUSING,
Figure 5.111b.

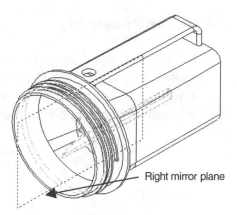

Right mirror plane

Figure 5.111a

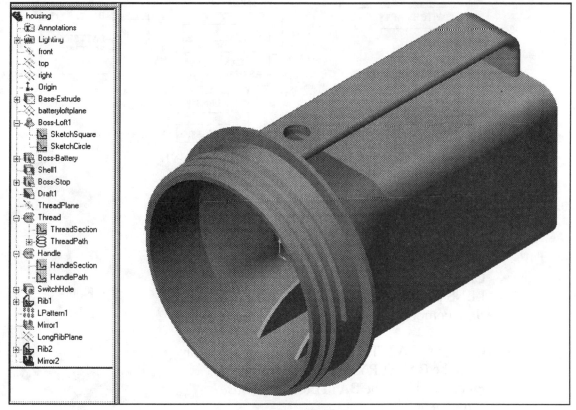

Figure 5.111b

The parts for the FLASHLIGHT are complete.

## 5.9    FLASHLIGHT Assembly

In Project 2, you created a simple assembly. You used the Bottom up design approach. Simplify the FLASHLIGHT assembly. Combine components into sub-assemblies. Identify the base component for each sub-assembly. Plan the sub-assembly component layout, Figure 5.112a.

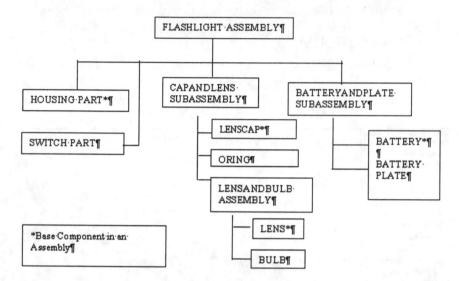

Component Layout Figure 5.112a

### 5.9.1    FLASHLIGHT Overview

The FLASHLIGHT assembly steps are as follows:

Figure 5-112b

- Create the LENSANDBULB sub-assembly from the LENS and BULB, Figure 5.112b. The LENS is the *Base* component.

- Create the BATTERYANDPLATE sub-assembly from the BATTERY and BATTERYPLATE, Figure 5.112c. The BATTERY is the *Base* component.

Figure 5.112c

- Create the CAPANDLENS sub-assembly from the LENSCAP, O-RING and LENSANDBULB sub-assembly. The LENSCAP is the *Base* component.

- Create the FLASHLIGHT assembly. The HOUSING is the B*ase* component, Figure 5.112d. Add the SWITCH, CAPANDLENS and BATTERYANDPLATE.

Figure 5.112d

- Modify the dimensions to complete the FLASHLIGHT assembly.

Close all parts before you begin this exercise. Click **Close All** from the Window menu.

### 5.9.2    Assembly Techniques

Assembly modeling requires practice and time. Below are a few helpful techniques to address the Bottom up modeling approach.

- Create an assembly layout structure. This will organize the sub-assemblies and components.

- Insert sub-assemblies and components as lightweight components. Lightweight components save on file size, rebuild time and complexity.

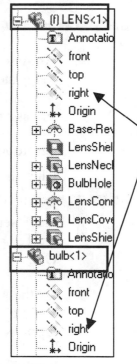

- Use the Zoom and Rotate commands to select the geometry in the mate process. Zoom to select the correct face. Filters are useful to select geometry.

- For clarity, apply different colors to features and components.

- Mate with reference planes when addressing complex geometry. Example: The O-RING does not contain a flat surface or edge.

- Activate Temporary axis and Planes from the View menu.

- Select reference planes from the FeatureManager. Expand the component to view the planes. Example: Select the *Right* plane of the LENS and the *Right* plane of the BULB to be collinear. Do not select the *Right* plane of the HOUSING if you want to create a mate between the LENS and the BULB, Figure 5.112e.

Figure 5-112e

- Remove display complexity. Hide components when not required. Suppress features when not required.

- Use Move Component, Rotate Component and Rotate Component around an Axis commands before mating. Position the component in the correct orientation.

- Conserve time.  Use Preview during the mate operation.  Hide unwanted features.  Create additional flexibility into a mate.  Use a distance mate with a zero value.  Flip the direction if required.

- Remove unwanted entries.  Use the Delete key from the Assembly Mating Items Selected text box.

- Verify the position of the mated components.  Use *Top*, *Front*, *Right* and *Section* views.

- Use caution when you view the color red in an assembly.  Red indicates that you are editing a part in the context of the assembly.

- Avoid unwanted references.

### 5.9.3   Create an Assembly Template

An Assembly Document Template is required to create the Flashlight Assembly and its sub-assemblies.

Create an assembly template. Click the **New** ☐ icon.  Click the **Assembly** Assembly icon from the Template dialog box.  Click **OK**.

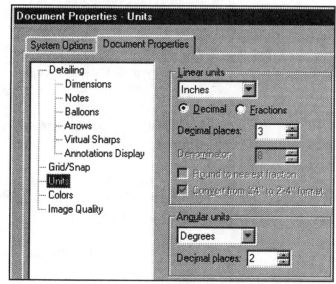

Figure 5.12f

Set the Assembly Document Template options.  Click **Tools**, **Options**, **Document Properties**.

Set the English units length increment.  Click the **Units** option.  Enter **Inches** from the Linear units list box.  Enter **3** in the Decimal places spin box, Figure 5.12f.

Rename *Plane1*, *Plane2* and *Plane3* to **Front**, **Top** and **Right**, respectively, Figure 5.12g.

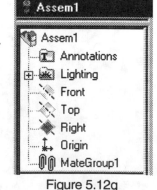

Figure 5.12g

Save the assembly template.  Click **File**, **SaveAs**.  Click the **\*.asmdot** from the Save As type list box.  Enter **ASSEMBLYENGLISHTEMPLATE** in the File name text box.  Click **Save**, Figure 5.12h.

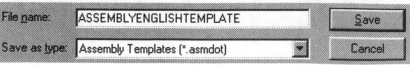

Figure 5.12h

### 5.9.4    LENSANDBULB Sub-assembly

Create the LENSANDBULB sub-assembly.  Click the **New** 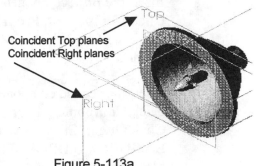 icon.  Click **ASSEMBLYENGLISHTEMPLATE**.  Click **OK**.

Click the **Save** 💾 icon.  Enter the sub-assembly name.  Enter **LENSANDBULB**.

Insert the *Base* component.  Click **Insert** from the Main menu.  Click **Component, From File**.  Enter **LENS**. Click **Open**.  Click the *Origin* in the LENSANDBULB Graphics window.

Coincident Top planes
Coincident Right planes

Insert the second component.  Click **Insert** from the Main menu.  Click **Component, From File**.  Enter **BULB**. Click **Open**.  Click **in front** of the LENS in the LENSANDBULB Graphics window, Figure 5.113a.

Figure 5-113a

Create the Mates between the BULB and LENS components:

1.  Create a **Coincident Mate**.  Click the ***Right*** plane of the LENS.  Click the ***Right*** plane of the BULB.  Click **Preview**.  Click **Apply**.

2.  Create a **Coincident Mate**.  Click the ***Top*** plane of the LENS.  Click the ***Top*** plane of the BULB.  Click **Preview**.  Click **Apply**.

Distance Mate

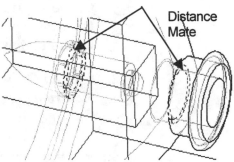

3.  Create a **Distance Mate**, Figure 5.113b. Click the **face** of the LENS.  Click the bottom **counterbore face** of the BULB. Enter **0.0** for Distance.  Click **Preview**.  Click **Apply**.

Figure 5.113b

The LENSANDBULB sub-assembly is fully defined,   Figure 5.114.

**Save** the LENSANDBULB  sub-assembly.

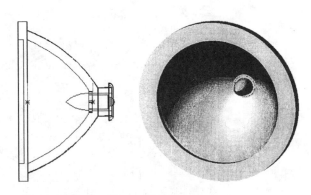

Figure 5.114

### 5.9.5　BATTERYANDPLATE Sub-assembly

Create the BATTERYANDPLATE sub-assembly. Click the **New**  icon. Click **ASSEMBLYENGLISHTEMPLATE**. Click **OK**.

Click the **Save** icon. Enter the sub-assembly name. Enter **BATTERYANDPLATE**. Click **Save**.

Insert the *Base* component. Click **Insert** from the Main menu. Click **Component, From File**. Enter **BATTERY**. Click **Open**. Click the *Origin* in the BATTERYANDPLATE Graphics window.

Insert the second component. Click **Insert** from the Main menu. Click **Component, From File**. Enter **BATTERYPLATE**. Click **Open**. Click above the BATTERY in the BATTERYANDPLATE **Graphics window**, Figure 5.115. Maximize the assembly window. Press the **f** key.

Figure 5.115

Create the Mates between the BATTERYPLATE and BATTERY components:

1.　Create a **Distance Mate**. Click the **bottom face** of the BATTERYPLATE. Click the **top narrow flat face** of the BATTERY. Enter **0.0**. Figure 5.115a. Click **Preview**. Click **Apply**.

2.　Create a **Coincident Mate**. Click the *Front* plane of the BATTERY. Click the *Front* plane of the BATTERYPLATE, Figure 5.115b. Click **Preview**. Click **Apply**.

3.　Create a **Concentric Mate**. Click the **circular face** *Terminal* feature of the BATTERY. Click the **circular face** *Holder* feature of the BATTERYPLATE. Click **Preview**. Click **Apply**.

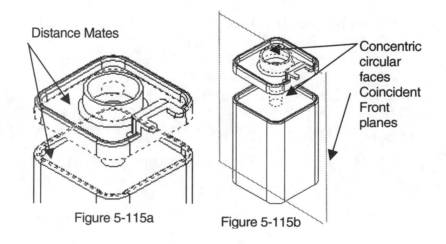

Distance Mates

Concentric circular faces
Coincident Front planes

Figure 5-115a　　　　Figure 5-115b

The BATTERYANDPLATE sub-assembly is fully defined, Figure 5-116.

**Save** the BATTERYANDPLATE.

### 5.9.6    CAPANDLENS Sub-assembly

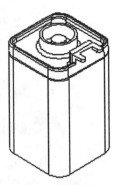

Create the CAPANDLENS sub-assembly.  Click the **New** ⬜ icon.  Click **ASSEMBLYENGLISHTEMPLATE**.

Click **OK**.  Click the **Save** 🖫 icon.  Enter the sub-assembly name.  Enter **CAPANDLENS**.  Click **Save**.

**Figure 5-116**

Insert the *Base* component.  Click **Insert** from the Main menu.  Click **Component, From File**.  Enter **LENSCAP**.  Click **Open**.  Click the *Origin* in the CAPANDLENS Graphics window.

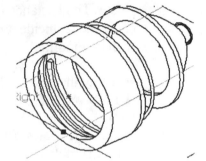

Insert the second component.  Click **Insert** from the Main menu.  Click **Component, From File**.  Enter **O-RING**.  Click **Open**.  Click **behind** the LENSCAP.

Insert the third component.  Click **Insert** from the Main menu.  Click **Component, From File**.  Enter **LENSANDBULB**.  Click **Open**.  Click **behind** the O-RING, Figure 5.117a.

**Figure 5.117a**

Caution: Select the correct reference.  Expand the LENSCAP and O-RING.  Click the *Right* plane within the LENSCAP.  Click the *Right* plane within the O-RING.  Do not select the *Right* plane from the top level FLASHLIGHT assembly.  You will create unwanted references.

Hide Components when not required for Mates.  Right-click on the **LENSANDBULB** subassembly in the FeatureManager.  Click **Hide**.

Create the Mates between the LENSCAP and O-RING component.  Select the planes from the FeatureManager.

1.  Create a **Coincident Mate**.  Click the *Right* plane of the LENSCAP.  Click the *Right* plane of the O-RING.  Click **Preview**.  Click **Apply**.

2.  Create a **Coincident Mate**.  Click the *Top* plane of the LENSCAP.  Click the *Top* plane of the O-RING.  Click **Preview**.  Click **Apply**.

3.  Create a **Distance Mate**.  Click the **back inside face** of the LENSCAP.  Click the *Front* plane of the O-RING, Figure 5.117b.  Enter **0.125/2** for Distance.

Click the *Right* view.  Click **Preview**.

If required, click the **Flip the dimension to the other side** check box.  Click **Apply**.

How is the Distance, 0.0625 calculated? Figure 5.117c.

Answer: O-RING Radius (.125"/2) = 0.0625".

Note: The Distance Mate option offers additional flexibility over the Coincident Mate option.  A Distance Mate value can be modified.

Display the LENSANDBULB.  Right-click **LENSANDBULB** in the FeatureManager.  Click **Show**.

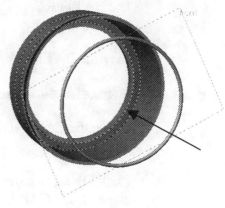

Figure 5.117b

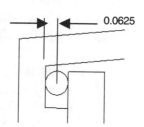

Figure 5.117c

Create the Mates between the LENSCAP and the LENSANDBULB sub-assembly.

1.  Create a **Coincident Mate**.  Click the *Right* plane of the LENSCAP.  Click the *Right* plane of the LENSANDBULB, Figure 5.117d.

2.  Create a **Coincident Mate**.  Click the *Top* plane of the LENSCAP.  Click the *Top* plane of the LENSANDBULB.

3.  Create a **Distance Mate**.  Click the **narrow inside face** of the LENSCAP.  Click the **front face** of the LENSANDBULB.  Enter **0.0** for Distance, Figure 5.117e.  Click **Preview**.  Flip the dimension to the other side to position the O-RING.  Click **Apply**.

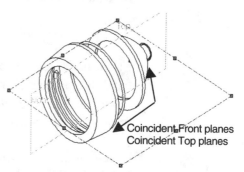

Figure 5.117d

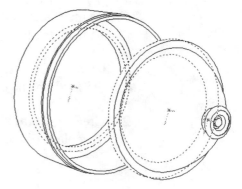

Figure 5.117e

The CAPANDLENS sub-assembly is fully defined. Confirm the location of the O-RING. Click the **Right** plane. Click **View** from the Main menu. Click **Display**. Click *Section*, Figure 5.118.

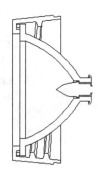

Figure 5.118

**Save** the CAPANDLENS.

### 5.9.7   Complete the Assembly

Create the FLASHLIGHT assembly. Click the **New** 🗋 icon. Click **ASSEMBLYENGLISHTEMPLATE**. Click **OK**.

Click the **Save** 🖫 icon. Enter **FLASHLIGHT**. Click **Save**.

Insert the *Base* component. Click **Insert** from the Main menu. Click **Component, From File**. Enter **HOUSING**. Click **Open**. Click the *Origin* of the FLASHLIGHT.

Figure 5.119

Insert the second component. Click **Insert** from the Main menu. Click **Component, From File**. Enter **SWITCH**. Click **Open**. Click **in front** of the HOUSING, Figure 5.119.

Create the Mates between the HOUSING and the SWITCH component:

1. Create a **Coincident Mate**. Click the **Right** plane of the HOUSING. Click the **Right** plane of the SWITCH. Click **Preview**. Click **Apply**.

2. Click **View** from the Main menu. Click **Temporary axis**. Create a **Coincident Mate**. Click the *temporary axis* inside the *Switch Hole* of the HOUSING. Click the *Front* plane of the SWITCH. Click **Preview**. Click **Apply**.

3. Create a **Distance Mate**, Figure 5.120. Click the **top face** of the *Handle*. Click the **Vertex** on the *Dome* of the SWITCH. Enter **0.200** for Distance. Click **Preview**. Click **Apply**.

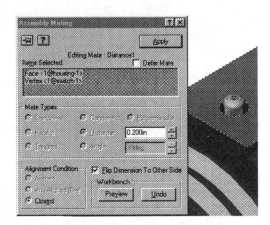

Figure 5.120

**Save** the FLASHLIGHT, Figure 5.121.

Insert the CAPANDLENS sub-assembly. Click **Insert** from
the Main menu. Click **Component**, **From File**. Enter
**CAPANDLENS**. Click **Open**. Click **in front** of the
HOUSING.

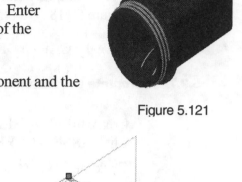

Figure 5.121

Create the Mates between the HOUSING component and the
CAPANDLENS sub-assembly:

1. Create a **Coincident Mate**,
   Figure 5.122. Click the *Right* plane
   of the HOUSING. Click the *Right*
   plane of the CAPANDLENS.

2. Create a **Coincident Mate**. Click
   the *Top* plane of the HOUSING.
   Click the *Top* plane of the
   CAPANDLENS.

3. Create a **Distance Mate**. Click the
   **front face of the *Boss-Stop*** on the
   HOUSING. Click the **back face** of
   the CAPANDLENS. Enter **0.0** for
   Distance, Figure 5.123.

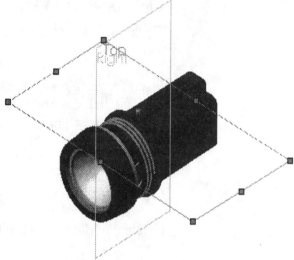

Figure 5.122

**Save** the FLASHLIGHT.

Figure 5.123

Insert the BATTERYANDPLATE sub-assembly. Click **Insert** from the Main menu. Click **Component, From File**. Enter **BATTERYANDPLATE**. Click **Open**. Click the FLASHLIGHT Graphics Window to the left of the HOUSING.

Rotate the part. Click the **Rotate Component around Axis** 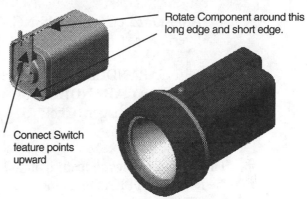 icon. Rotate the BATTERYANDBATTERYPLATE until the *ConnectorSwitch* feature points vertical, Figure 5.124.

Create the Mates between the HOUSING component and the BATTERYANDPLATE sub-assembly:

1.  Create a **Coincident Mate**. Click the *Right* plane of the HOUSING. Click the *Front* plane of the BATTERYANDPLATE. Click **Preview**. Click **Apply**.

Rotate Component around this long edge and short edge.

Connect Switch feature points upward

Figure 5.124

2.  Create a **Coincident Mate**. Click the *Top* plane of the HOUSING. Click the *Right* plane of the BATTERYANDPLATE. Click **Preview**. Click **Apply**.

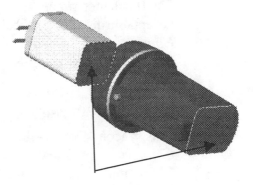

3.  Create a **Distance Mate**, Figure 5.125. Click the **back face** of the HOUSING. Click the **bottom face** of the BATTERYANDPLATE. Enter **0.275** for Distance. Click **Flip** if required. Click **Preview**. Click **Apply**.

Figure 5.125

**Rebuild** the FLASHLIGHT, Figure 5.126.

**Save** the FLASHLIGHT.

Click **Yes** to update all components.

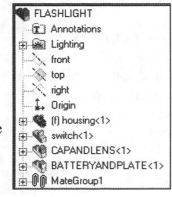

- FLASHLIGHT
  - Annotations
  - Lighting
  - front
  - top
  - right
  - Origin
  - (f) housing<1>
  - switch<1>
  - CAPANDLENS<1>
  - BATTERYANDPLATE<1>
  - MateGroup1

Figure 5.126

### 5.9.8  Addressing Design Issues

There are interference issues with the assembly, Figure 5.127a.  Address these design issues:

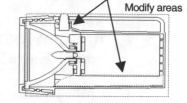

Figure 5.127a

- Reduce the *SwitchConnector* feature size on the BATTERYPLATE

- Adjust *Rib2* on the HOUSING.  Test with the Interference Check command.

Hide the CAPANDLENS.  Right click on **CAPANDLENS** in the FeatureManager.  Click **Hide Component**.

The *ConnectorSwitch* feature of the BATTERYPLATE is too long.  Contain the *ConnectorSwitch* within the HOUSING.  Use the Measure function to determine an acceptable distance between the *ConnectorSwitch* feature of the BATTERYPLATE, Figure 5.127b.

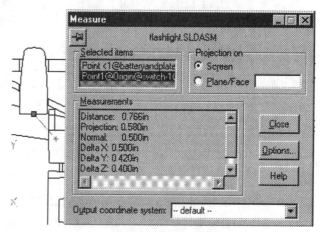

Figure 5.127b

Expand the BATTERYPLATE.  Click the **Plus** ⊞ icon.  Double-click the *ConnectorSwtich* feature.  Double click the **1.000** dimension.  Enter **0.500**.  Click **Rebuild**, Figure 5.127c.

Figure 5.127c

There is an interference fit issue between the HOUSING and the BATTERY.  Click **Tools** from the Main menu.  Click **Interference Check**.  The Interference Volumes dialog box is displayed.  Click the **BATTERY**.  Click the **HOUSING**.  Click the **Check** button.  The interference is displayed in the Graphics window.  *Rib2* overlaps the BATTERY by 0.050", Figure 5.127d.

Modify *Rib2* by 0.050".  Expand the **HOUSING** in the FeatureManager.  Double-click on the *Rib2* feature.

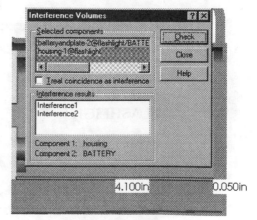

Figure 5.127d

Double click on **1.300**.  Enter **1.350**.  Click **Rebuild**, Figure 5.127e.

Note: Interference must exist
between the BULB and the
BATTERY to create an electrical
connection.

Show the CAPANDLENS.  Right-
click on **CAPANDLENS**.  Click
**Show**.  **Show** all hidden components
and features.  The FLASHLIGHT
design is complete.

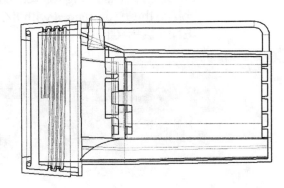

Figure 5.127e

**Save** the FLASHLIGHT.  Click
**YES** to the question, "Rebuild the assembly and update the components".

## 5.10  Export Files

You receive a call from sales.  They inform you that the customer increased the
order to 200,000 units.  However, the customer requires a prototype to verify the
design in six days.  What do you do?  Answer: Contact a Rapid Prototype
supplier.

You export three SolidWorks files:

- HOUSING

- LENSCAP

- BATTERYPLATE

Use the Stereo Lithography (STL) format.  Email the three files to a Rapid
Prototype supplier.

Example: Paperless Parts Inc. (www.paperlessparts.com).

A Stereolithography (SLA) supplier provides physical models from 3D drawings.
2D drawings are not required.

Open the LENSCAP.  Expand the **CAPANDLENS** sub-assembly.  Right-click on
the **LENSCAP** from the FLASHLIGHT FeatureManager.  Click **Open
LENSCAP.SLDPRT**.

Export the LENSCAP.  Click **File**, **SaveAs**.  A warning message states that "Lens-cap.SLDPRT is being referenced by other open documents.  "Save AS" will replace these references with the new name", Figure 5.128.  Create a copy of LENSCAP.  Click **OK**.

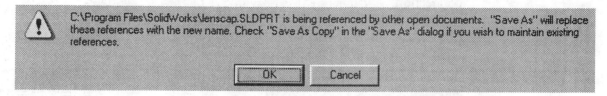

Figure 5.128

Click the **Save as copy** check box from the Save dialog box, Figure 5.129.

Click **STL Files (\*.stl)** from the Save as Type drop down list, Figure 5.130a.  The Save dialog box displays new options, Figure 5.130b.

Figure 5.129

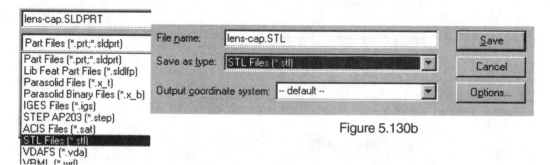

Figure 5.130b

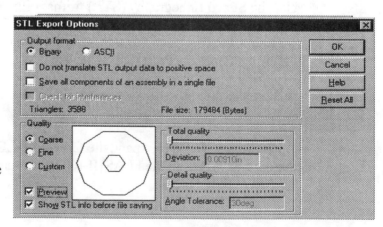

Figure 5.130a

Click the **Options** button in the lower right corner of the Save dialog box.  The STL Export Options dialog is displayed, Figure 5.131.

Click **Binary** from the Output format box.  Click **Course** from the Quality box.

Figure 5.131

Display the STL triangular faceted model, Figure 5.132. Click **Preview**.

Create the binary STL file. Click **OK** from the STL dialog box. Click **Save** from the Save dialog box. A status report is provided, Figure 5.133. Save the file LENS-CAP.STL. Click **Yes**.

Figure 5.132

Figure 5.133

The STL file created in binary format is 179484 Bytcs. The STL file created in ASCII format is 1,015,428 Bytes, Figure 5.134.

You receive the three SLA physical models for the supplier.

Figure 5.134

You assemble the rapid prototype models with purchased components.

Results:

A flashlight assembly delivered to the customer in six days. Success!

It is time to go home. The telephone rings. Guess who? The customer is ready to place the order. Tomorrow you will receive the purchase order. You think about the concerns of manufacturing, purchasing and shipping. In the rush to create the prototype, you forgot about the packaging, the part numbers and the silk screen vendor. Congratulation, the project is just beginning!

Let's try some more examples.

**5.11 Questions**

1.  Identify the function of the following features:

    *   Sweep

    *   Revolved Thin Cut

    *   Loft

    *   Rib

    *   Pattern

2.  What are suppressed features

3.  Why would you suppress a feature?

4.  The Rib features require a Sketch, thickness and a _____ direction.

5.  What is a Pierce relation?

6.  What is the advantage of the Distance Mate option over the Coincident Mate option

7.  How do you create a thread using the Sweep feature?

8.  How do you create a Linear Pattern?

9.  Provide 5 proven assembly modeling techniques.

10. How do you determine interference between components in an assembly?

## 5.12 Exercises

I. Create O-RINGs with various cross sections using the *Base-Sweep* feature.

Exercise 5.1          C-SHAPE-O-RING

Exercise 5.2          QUATTRO-SEAL-O-RING

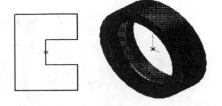

Exercise 5.1 C-Shaped O-RING

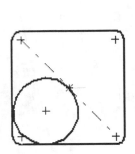

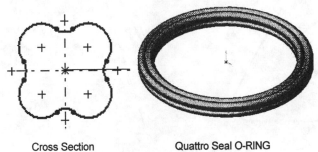

Cross Section

Quattro Seal O-RING
Exercise 5.2

Exercise 5.3          Create the OFFSET-LOFT

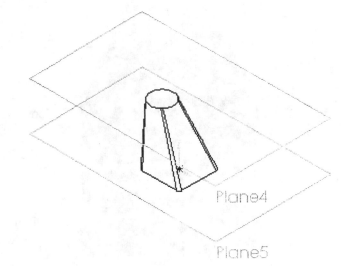

II.        A plastic BOTTLE is created from a variety of SolidWorks features. Create the shoulder of the BOTTLE with the *Loft* Base feature.

Exercise 5.4: Triangular Shaped Bottle

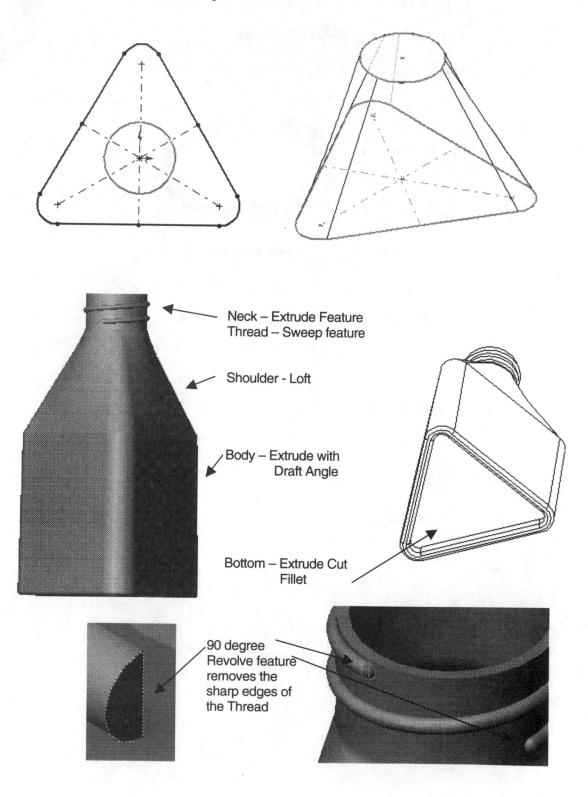

Neck – Extrude Feature
Thread – Sweep feature

Shoulder - Loft

Body – Extrude with
Draft Angle

Bottom – Extrude Cut
Fillet

90 degree
Revolve feature
removes the
sharp edges of
the Thread

Exercise 5.5: Oval Shaped Bottle          Exercise 5.6: Rectangular Shaped Bottle

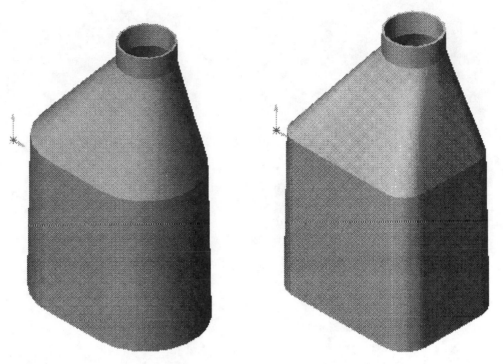

Exercise 5.7: Rectangular Shaped Bottle with Handle

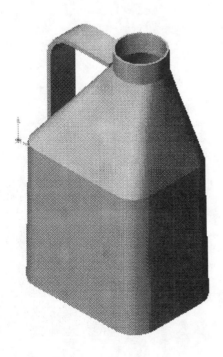

**NOTES:**

# Project 6

## Top Down Assembly

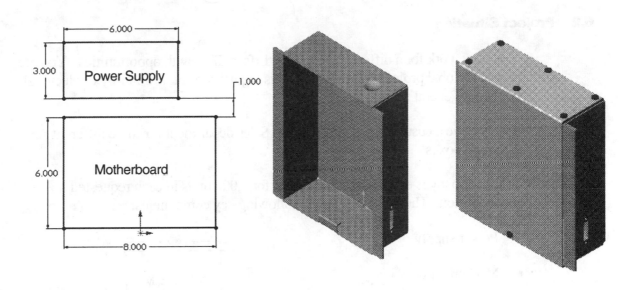

Below are the desired outcomes and usage competencies based upon completion of Project 6.

| Project Desired Outcomes | Usage Competencies |
|---|---|
| • Knowledge and experience in Top down assembly creation. | • Ability to create a new Top down assembly and layout sketch. |
| • Understanding of components created in the context of the assembly. | • Knowledge to create and modify components developed in the context of the assembly. |
| • Comprehension of sheet metal design basics. | • Understanding of sheet metal Bend and Rip features. |
| • Family of electrical boxes. | |

# 6    Project 6 – Top Down Assembly Modeling

## 6.1    Project Objective

Create a family of electrical boxes using the Top down assembly modeling approach.

## 6.2    Project Situation

You now work for a different company. Life is filled with opportunities. You are part of a global project design team that is required to create a family of electrical boxes for general industrial use.

You receive a customer request from the Sales department for three different size electrical boxes.

Your company is expecting a sales order for 5,000 units in each requested configuration. The box contains the following key components:

- Power supply

- Motherboard

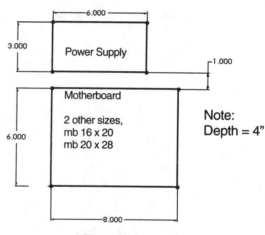

Figure 6.1a

The size of the power supply is the same for all boxes. There are three different motherboard sizes, Figure 6.1a. In other words, the size of the power supply is the constant and the size of the motherboard is the variable. Three different BOX sizes are determined by the physical design constraint of the motherboard. The BOX is constructed of aluminum.

You contact the customer to discuss and obtain design options and product specifications. Key customer requirements:

- Three different BOX sizes:

Physical design constraint of the motherboards:

1.  8" x 6" x 4"          Small

2.  16" x 20" x 4"        Medium

3.  20" x 28" x 4"        Large

- Adequate spacing between the power supply, motherboard and internal walls.

- Field serviceable.

You are responsible to produce a sketched layout from the provided critical dimensions. You are also required to design the outside sheet metal BOX. Note: The BOX is used in an outside environment.

## 6.3   Top Down Design Approach

The Top down design approach is a conceptual approach used to develop products from within the assembly. The major design requirements are translated into sub-assemblies or individual components and key relationships, Figure 6.1b.

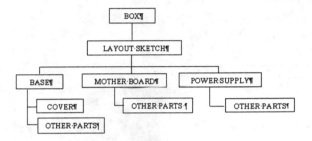

Figure 6.1b

Start with a *Layout Sketch*. The *Layout Sketch* specifies the location of the key components. Create or add additional components to complete the BOX.

You use a combination of Top down and Bottom up approaches in the BOX assembly.

Consider the following in a preliminary design product specification:

- What are the major components in the design? The motherboard and power supply are the major components.

- What are the key design constraints? Three motherboard sizes and the power supply size are specified by the customer.

- How does each part relate to the other? From past experience and discussions with the electrical engineering department, a 1" physical gap is required between the power supply and the motherboard.

- How will the customer use the product? The customer does not disclose the specific usage of the electrical boxes.

- What is the most cost-effective material for the product? Aluminum is cost-effective, strong, easy to fabricate, corrosion resistant, non-magnetic and easily finished.

Create the first BOX. The BOX dimension is 8" x 6" x 4". Verify the design feasibility. Then create the other two boxes through the SolidWorks design table.

The Top down design approach is developed in this Project. Sheet metal components are used as an example to illustrate the design approach.

## 6.4    Project Overview

Create the BOX *Layout Sketch* in the assembly, Figure 6.2a. Use *Equations* to insure BOX integrity for size changes.

Create the first sheet metal component. The first sheet metal component is the BASE, Figure 6.2b. Create the flat manufactured state. Add sheet metal bends and additional features, Figure 6.2c.

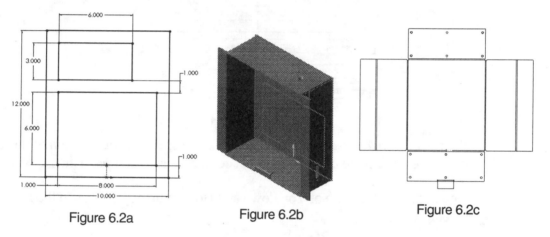

Figure 6.2a                          Figure 6.2b                          Figure 6.2c

Create the second sheet metal component. The COVER is the second component, Figure 6.3a.

Work with linear patterns of holes. Create a linear pattern for the BASE and COVER. Reference the geometry and use *Equations* to pattern holes and screws in the assembly, Figure 6.3b.

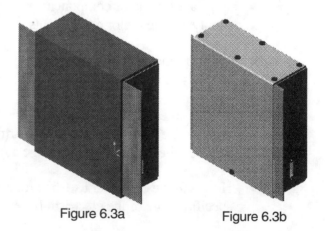

Figure 6.3a                          Figure 6.3b

Use a design table to create the other two boxes, Figure 6.3c.

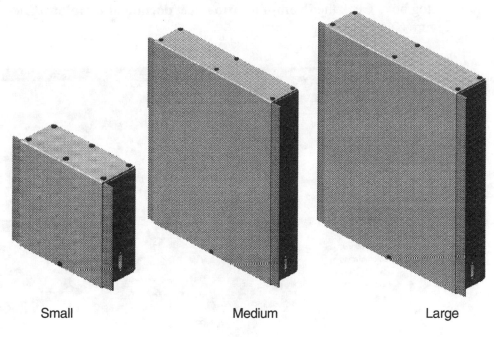

| Small | Medium | Large |

Figure 6.3c

## 6.5    Layout Sketch

Utilize the *Layout Sketch* to develop component space allocations and relations. Components and sub-assemblies reference the *Layout Sketch*.

The BOX contains the following key components:

- Power supply

- Motherboard

The minimum physical spatial gap between the motherboard and the power supply is 1".

The minimum physical spatial gap between the motherboard, power supply and the internal sheet metal BOX wall is 1".

Create the BOX.  Click the **New** icon.

Click the **ASSEMBLYENGLISHTEMPLATE** .  Click **OK**.

Click the **Save** icon.  Enter **BOX**.  Click **SAVE**.

In this project use an ENGLISH TEMPLATE for all parts, assemblies and drawings. Click **Tools**, **Options**. Click **Default Templates** from the Systems dialog box. Click the **Prompt user to select document template** dialog box, Figure 6.4a.

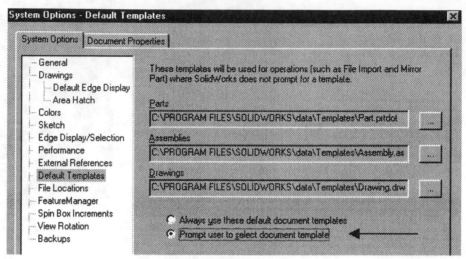

Figure 6.4b

The system displays the new template dialog box when a new part is inserted into an assembly.

Click the **General** option. Uncheck the **Edit design tables in a separate window** box. Uncheck **Enable PropertyManager** box, Figure 6.4b.

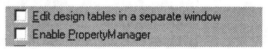

Figure 6.4b

Create a *Layout* S*ketch* in the assembly. Select the *Sketch* plane. The *Front* plane is the default *Sketch* plane.

Sketch the *Layout*. Click the **Sketch** icon. Click the **Front** icon.

Sketch the profile of the motherboard. Click the **Rectangle** icon. Position the **first point** of the rectangle away from the *Origin* and to the left of the *Right* plane, Figure 6.5a.

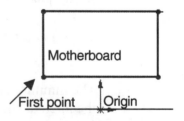

Figure 6.5a

Sketch the profile of the power supply. The mouse pointer displays the rectangle. The Rectangle command is currently selected. Click the **first point** of the rectangle to the left edge of the motherboard, Figure 6.5b.

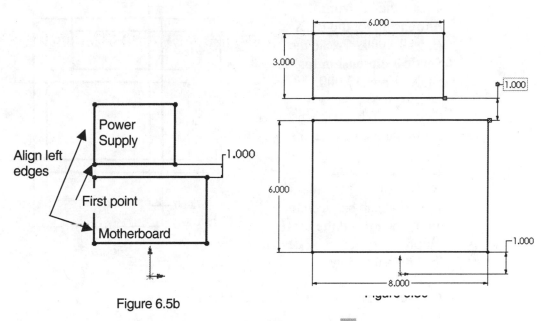

Align left edges

Power Supply

First point

Motherboard

Figure 6.5b

Create a collinear relation. Click the **Add Relations** icon. Click the **left vertical line of the motherboard** profile. Click the **left vertical line of the power supply**. Click **Collinear**. Click **Apply**. Click **Close**.

Add a **vertical dimension** between the *Origin* and the motherboard. Enter **1.000**. Add a **vertical dimension** between the motherboard and the power supply. Enter **1.000**. Dimension the **motherboard** profile. Dimension the **power supply** profile, Figure 6.5c.

Sketch the outside profile of the **BOX**, Figure 6.6a. Click the **Rectangle** icon. Position the **first point** of the BOX to the left of the *Origin* and collinear with the *Top* plane. Position the **second point** of the BOX to the right of the power supply.

Add a midpoint relation. Click the **Add Relations** icon. Right-click in the **Graphics window**. Click **Clear Selection**. Click the *Origin*. Click the **bottom horizontal line**. Click **Mid point**. Click **Apply**. Click **Close**.

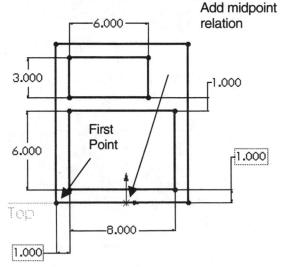

Add midpoint relation

First Point

Figure 6.6a

Add dimensions. The BOX must enclose the motherboard, power supply and maintain a 1.000" spatial gap, Figure 6.6b. Create a **horizontal dimension**. Enter **1.000**.

Create the **horizontal dimension** for the BOX. Enter **10.000**. Create the **vertical dimension** for the BOX. Enter **12.000**.

Question: What happens when the motherboard size changes?

How can you insure that the BOX maintains the required 1" spatial gap between the motherboard and the power supply? How can you design for future revisions?

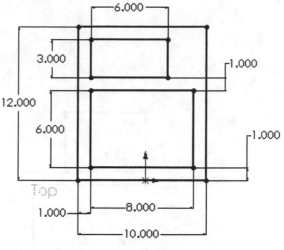

Figure 6.6b

Answer: Through Link Values and *Equations*.

### 6.5.1  Link Values and Equations

Link Values are used to define equal relations. Create an equal relation between two or more sketched dimensions and or features with a Link Value. Link Values require a shared name.

Mathematical expressions that define relationships between parameters and or dimensions are called *Equations*.

*Equations* use shared names to control dimensions. Use *Equations* to connect values from sketches, features, patterns and various parts in an assembly.

Use Link Values within the same part. Use *Equations* in different parts and assemblies.

The project goal in this section is to create a family of different size boxes; namely three.

Insure that the models remain valid when dimensions change for different internal components. This is key! Create a Link Value.

Right click on the lower vertical dimension **1.000**, Figure 6.7a. Click **Link Values**.

The Shared Values dialog box is displayed, Figure 6.7b. Enter **gap** in the Name text box. Gap is the Link Value name. Click **OK**.

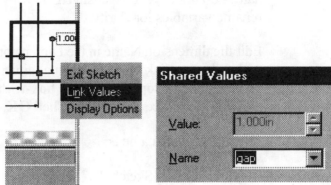

Link the remaining 1.000 dimensions, Figure 6.8. Right click on the upper vertical dimension **1.000**. The vertical dimension is located between the power supply and the motherboard. Click **Link Values**. The Shared Values dialog box is displayed. Click the **drop down arrow** from the Name text box. Click **gap**.

Figure 6.7a                      Figure 6.7b

Link the horizontal **1.000** dimension.

Test the Link Values. Double-click the **1.000** dimension between the power supply and the motherboard. Enter **0.500**.

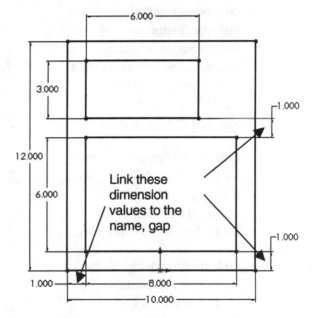

Link these dimension values to the name, gap

Figure 6.8

Click the **Checkmark** ✓ icon in the Modify dialog box. The two Link Values change. Return to the original value. Click the horizontal dimension, **0.500**. Enter **1.000**. Click the **Checkmark** ✓ icon. All Link Values are equal to 1.000.

Each dimension has a unique variable name. The names are used as *Equation* variables. The default names are based on the *Sketch*, *Feature* or *Part*. Feature names do not have to be changed. However, when creating numerous equations, rename variables for clarity.

Edit the dimension Name in the Dimension Properties dialog box for overall box width. Right-click on the horizontal dimension, **10.000**. Click **Properties**. Enter **box-width** in the Name text box, Figure 6.9. Click **OK**.

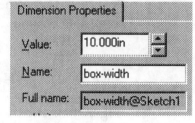

The full variable name is:

  "box-width@Sketch1"

Figure 6.9

Note: The system automatically appends Sketch1. If features are created or deleted in a different order, your variable names will be different. Select the correct dimensions. Follow the variable names and location in Figure 6.10.

Edit the dimension Name for the overall box height. Right-click on **12.000**. Enter **box-height** in the Name text box. Click **OK**.

Edit the dimension Name for the motherboard width. Right-click on **8.000**. Enter **mb-width** in the Name text box. Click **OK**.

Note: Use mb as the abbreviation for motherboard.

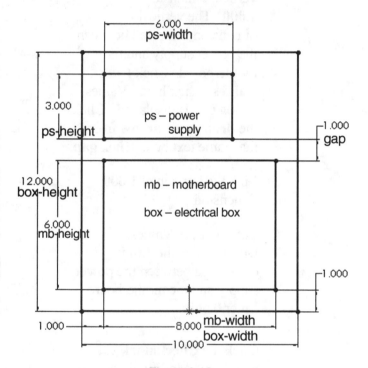

Figure 6.10

Edit the dimension name for the motherboard height. Right-click on **6.000**. Enter **mb-height** in the Name text box. Click **OK**. Edit the dimension name for the power supply width. Right-click on **6.000**. Enter **ps-width** in the Name text box.

Note: Use ps as the abbreviation for power supply. Click **OK**. Edit the dimension name for the power supply height. Right-click on **3.000**. Enter **ps-height** in the Name text box. Click **OK**.

The first *Equation* defines the width of the box. The box width is dependent upon the dimension of the motherboard and the required gap between the motherboard and the side walls of the BOX.

Create the first *Equation*. Click the
**Select** icon. Click the horizontal
box-width dimension, **10.000**. Click
**Tools** from the Main menu. Click
*Equations*. The Equations dialog box
appears, Figure 6.11a.

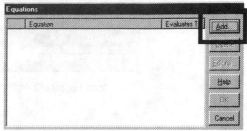

Figure 6.11a

Add a new *Equation* to the text box. Click **Add** from the Equations dialog box. The New Equation dialog box appears, Figure 6.11b.

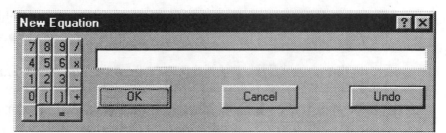

Figure 6.11b

The variable name, "box-width@Sketch1" is followed by an equal sign =. This is added to the Equation text box, Figure 6.11c.

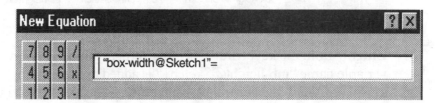

Figure 6.11c

Click the horizontal mb-width dimension, **8.000**. The variable name "mb-width@Sketch1" is added to the text box, Figure 6.11d.

**Enter + 2 \*** from the keyboard. Click the gap horizontal dimension, **1.000**. The variable name, "gap@Sketch1" is added to the text box.

Figure 6.11d

Display the first *Equation*. Click **OK** from the New Equation dialog box. The Equations dialog box contains the complete *Equation*, Figure 6.12.

Figure 6.12

A green check mark √, placed in the first column indicates that the *Equation* is solved. The number # represents the *Equation* number. Example: *Equation 1*. The value 10 in the Evaluates To column represents the result of the solved *Equation*.

The second *Equation* defines the height of the BOX. The height of the BOX is dependent upon the height of the motherboard and power supply.

Create the second *Equation*. Click **Add** from the Equations dialog box. The New Equation dialog box appears. Click the vertical box-height dimension, **12**. The variable name, "box-height@Sketch1" is added to the Equation text box, Figure 6.13.

Figure 6.13

**Enter** = from the keyboard. Click the vertical mb-height dimension, **6.000**. The variable name "mb-height@Sketch1" is added to the text box. **Enter** + from the keyboard. Click the vertical ps-height dimension, **3.000**. The variable name "ps-height@Sketch1" is added to the text box.

**Enter + 3 \*** from the keyboard.

Click the gap vertical dimension, **1.000**. The variable name, "gap@Sketch1" is added to the text box.

Display the second *Equation*. Click **OK** from the New Equation dialog box.

The number 12 in the Evaluates To column represents the result of the solved *Equation*. Drag the **slider bar** to display the *Equation*, Figure 6.14.

Figure 6.14

The two *Equations* are complete. Click **OK** from the Equations text box. The *Equations* folder appears below the *Origin* in the FeatureManager.

Close the *Layout Sketch*. Click the **Sketch** icon.

Rename *Sketch1* to *LayoutSketch*, Figure 6.15.

Display all dimensions. Right-click on the *Annotations* folder. Click **Show Feature Dimensions**. Edit the *Equations*. Right-click on the *Equations* folder. Click **Edit**.

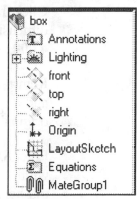

Figure 6.15

Note: The variable names are updated to include the new sketch name *LayoutSketch*, Figure 6.16. Click **OK**.

Figure 6.16

*Equation 1* drives the variable, box-width. Do not modify the box-width. The box-width is the dependent variable.

Change the dimensions to verify the *Equations*. Verify the *Equations* for values greater than the original value. Click the mb-width dimension, **8.000**. Enter **16.000**. Click

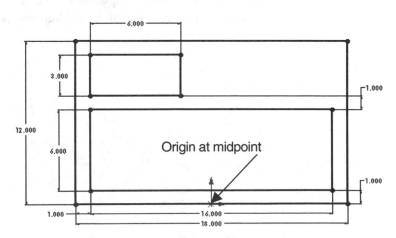

Figure 6.17

**Rebuild**, Figure 6.17. The *Origin* is positioned at the midpoint of the bottom edge.

The box-width is dependent upon the variables on the right hand side of the *Equation*:

- mb-width

- gap

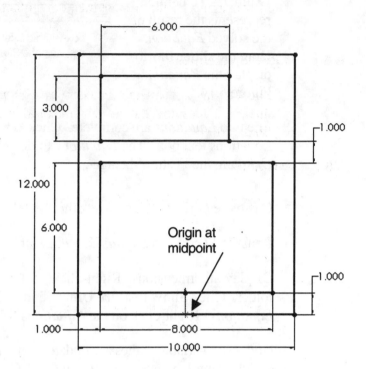

Click the mb-height dimension, **6.000**.  Enter **4.000**.  Click **Rebuild**.

Return the mb-width and mb-height to the original values.

Enter **8.000** for mb-width.  Enter **6.000** for mb-height, Figure 6.18.  Click **Rebuild**.

Figure 6.18

Hide all dimensions.  Right-click the ***Annotations*** folder.  Remove the checkmark. Click **Show Feature Dimensions**.

**Save** the BOX assembly.  The *Layout Sketch* is complete.

Note: When you save the assembly and exit SolidWorks, save both the assembly and the referenced components.  When you return to SolidWorks, open the assembly before opening individual components referenced by the assembly.

## 6.6 Sheet Metal Overview

You need to understand some basic sheet metal definitions before starting the BOX. Talk to colleagues. Talk to sheet metal manufacturers. Review other sheet metal parts previously designed by your company.

### 6.6.1 Material Thickness

Sheet metal parts are fabricated from a flat piece of raw material. The material thickness does not change. The material is cut, formed and folded to produce the final part.

### 6.6.2 Design State

There are two design states for sheet metal parts:

- Bent

- Flat

Work in the bent 3D state and then flatten the part or work in the flat 2D state and then add individual bend lines and cuts.

### 6.6.3 Neutral Bend Line

Example: Use a flexible eraser, 2" or longer. Bend the eraser in a U shape. The eraser displays tension and compression forces. The area where there is no compression or tension is called the neutral axis or neutral bend line, Figure 6.19.

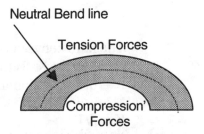

Neutral Bend line

Tension Forces

Compression' Forces

Figure 6.19

### 6.6.4 Developed Length

Assume the material has no thickness. The length of the material formed into a 360° circle is the same as its circumference. The length of a 90° bend would be ¼ of the circumference of a circle, Figure 6.20.

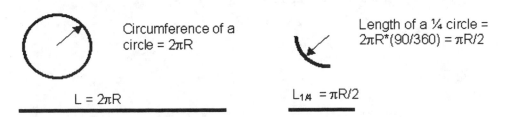

Circumference of a circle = $2\pi R$

$L = 2\pi R$

Length of a ¼ circle = $2\pi R*(90/360) = \pi R/2$

$L_{1/4} = \pi R/2$

Figure 6.20

In the real world, materials do have thickness. Materials develop different lengths when formed in a bend depending on their thickness. There are three major properties which determines the length of a bend:

1. Bend radius

2. Material thickness

3. Bend angle

The distance from the inside radius of the bend to the neutral bend line is labeled, $\delta$, Figure 6.21a. The symbol '$\delta$' is the Greek letter, delta. The amount of flat material required to create a bend is greater than the inside radius and depends upon the neutral bend line. The true developed flat length, L, is measured from the endpoints of the neutral bend line.

Example:

Material thickness, T = 0.100". Create a 90° bend with an inside radius of R = 0.150".

$$L = \tfrac{1}{2}\,\pi R + \delta T$$

The ratio between $\delta$ and T is called the K-factor. Let K = 0.41 for Aluminum.

$$K = \delta / T$$

$$\delta = KT = 0.041$$

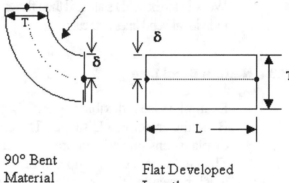

90° Bent Material

Flat Developed Length

Figure 6.21a

$$L = \tfrac{1}{2}\,\pi(0.150") + (0.041")(0.100") = 0.2397"$$

Note: Every shop uses their own numbers. For example one shop uses K = 0.41 for Aluminum. Another shop uses K = 0.45. Use tables, manufactures specifications or experience.

### 6.6.5   Relief

Sheet metal corners are subject to stress.
Excess stress will tear material.  Remove
material to relieve stress, Figure 6.21b.

## 6.7   Mate a New Component to the Layout Sketch

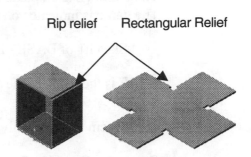

The BASE component is the first part in
the assembly.  Create the BASE
component inside the assembly and
attached it to a plane.  The BASE
component references the *Layout Sketch*.  First, create the BASE component as a
solid part.  Second, add the *Rip* feature and *Insert Bend* feature to create a sheet
metal part.

Figure 6.21b

Note: Solid components, sheet metal components or a combination of solid and
sheet metal components utilize a *Layout Sketch*.

Display the *Layout Sketch* in an *Isometric* view.  Click the **Isometric** icon.
Click the *Layout Sketch*.

Create the BASE component.  Click
**Insert** from the Main menu.  Click
**Component**.  Click **New Part**.

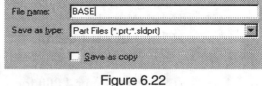

The Save As dialog box is displayed.
Enter **BASE** in the Filename text box,
Figure 6.22.  Click **Save**.

Figure 6.22

Components added in the context of
the assembly automatically receive an
In Place Mate within MateGroup1.

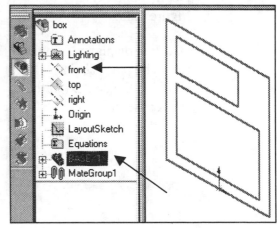

Select the In Place Mate plane.  Click
the *Front* plane of the BOX in the
FeatureManager.  The *Front* plane of
the component is mated with the
*Front* plane of the BOX, Figure 6.23.

The BASE is added to the
FeatureManager.  The system

Figure 6.23

automatically selects the Edit Part
icon.  The BASE text appears in the FeatureManager.  The BASE text is displayed
in light red to indicate that the part is actively being edited.

The current *Sketch* plane is the *Front* plane. The current sketch name is *Sketch1*. The name is indicated on the current graphics window title,

> "Sketch1 of BASE - in- BOX."

BASE is the name of the component created in the context of the assembly BOX.

The system automatically selects the Sketch icon.

Create the *Extruded-Base*
feature for the BASE.

Define the boundary of the
BOX. Extract existing edges
from the *Layout Sketch*.

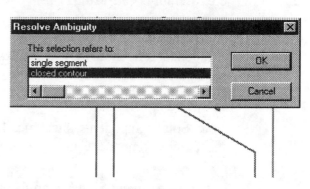

Create the *Sketch*. Click the
**right vertical line**,
Figure 6.24. Click the
**Convert Entities** icon.

Figure 6.24

The Resolve Ambiguity
dialog box is displayed.
Click **closed contour**.
Click **OK**. Extrude the
*Sketch*. Click the **Extrude**
icon. Enter **4.000** for
Depth. Click **OK**,
Figure 6.25.

BASE is displayed in light
red. The part is being
edited in the context of the
BOX assembly.

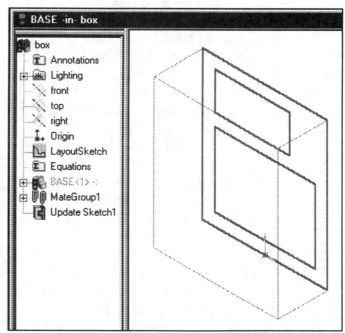

Figure 6.25

Display the *Extruded-Base* feature. **Expand** the BASE part 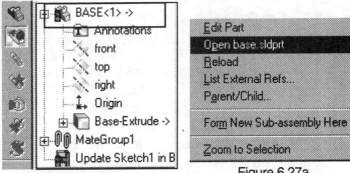 icon in the FeatureManager, Figure 6.26.

The features are displayed in gray. A box is drawn around the BASE to indicate the in-context part.

The *Extruded-Base* feature is complete. No other geometry is required from the *LayoutSketch* or BOX.

Note: Create additional features directly in the part.

Figure 6.26

Figure 6.27a

Return to the assembly.

Click the **Edit Part** icon. The BASE and its features are displayed in black.

**Save** the BOX assembly. Click **Yes** to Rebuild the models. Click **Yes** to save the referenced models.

Right-click **BASE** from the FeatureManager. Click **Open-BASE.sldprt**, Figure 6.27a.

The BASE is opened. The *Front* view is displayed. Display the *Isometric* view.

Click the **Isometric** icon, Figure 6.27b.

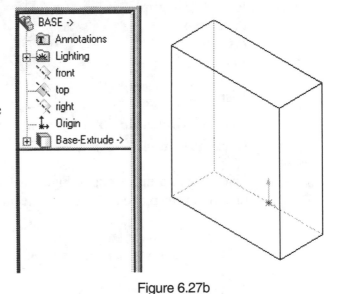

Figure 6.27b

Note: Do not create unwanted geometry references. Open the part when creating features that require no references from the assembly. The features of the BASE require no additional references from the BOX assembly.

## 6.8   Create the Shell Feature

Sheet metal features for the BASE are created from a solid extrusion. The *Shell* feature removes faces from the solid and creates the sheet metal thickness.

Create the *Shell*. Click the **front face** of the *Extruded-Base* feature, Figure 6.28a. Click the **Shell** 🔲 icon. Enter **0.100** for Thickness. Click **OK**, Figure 6.28b.

### 6.8.1 Create the Rip Feature

The *Rip* feature creates a cut along the edges of the *Extruded-Base* feature. *Rip* the *Extruded-Base* feature along the four edges.

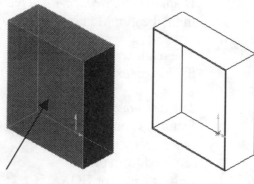

Figure 6.28a           Figure 6.28b

Display the Sheet Metal toolbar. Click **View**, **Toolbars**, **Sheet Metal**.

Create the *Rip*. Click the **Rip** 🔲 icon from the Sheet Metal toolbar. The Rip dialog box is displayed, Figure 6.29.

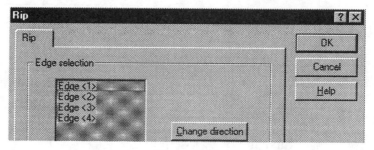

Figure 6.29

Display the part.

Click the **Hidden In Gray** 🔲 icon. Rotate the part. Click the **Rotate** 🔲 icon.

Select the four inside edges, Figure 6.30a. The arrow direction is displayed. Accept the default arrow direction.

Rip the 4 inside edges

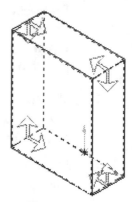

- Click the **lower left inside edge**

- Click the **upper left inside edge**

- Click the **upper right inside edge**

- Click the **lower right inside edge**

Figure 6.30a

Display the four *Rips*. Click **OK**, Figure 6.30b.

*Rip1* is added to the FeatureManager.

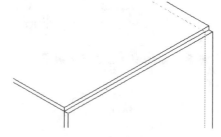

Figure 6.30b

**Save** the BASE.

### 6.8.2  Insert Sheet Metal Bends

The *Bend* feature creates sheet metal bends.  Specify bend parameters such as bend radius, bend allowance and relief.

Select the bottom face to remain fixed during bending and unbending.  Click the inside **bottom face**, Figure 6.31.

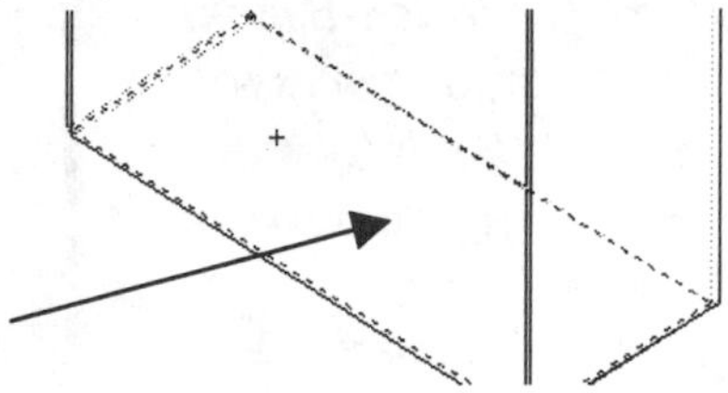

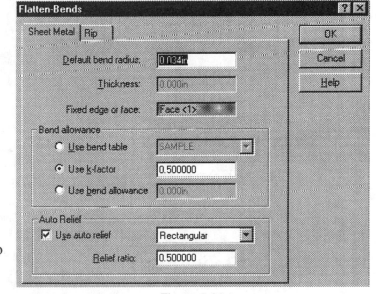

Figure 6.31

Click the **Insert Bend**  icon from the Sheet Metal toolbar.  The Flatten Bends dialog box appears, Figure 6.32.

Accept the Default bend radius, **0.034**. Accept the "Use k-factor".  Click the **Use auto relief** check box. Enter **Rectangle** from the Auto Relief drop down list.

Display the bends. Click **OK** to the system message, "Auto Relief cuts were made for one or more bends".

Figure 6.32

Figure 6.33

Display the *Bends*.  Click **Rotate**.  Click **Zoom**, Figure 6.33.

Display the flat state.

Drag the **Rollback** bar upward between *Flatten-Bends1* and *Process-Bends1*, Figure 6.34.

The part is displayed in its flat state, Figure 6.35.

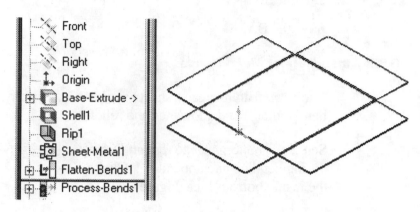

Figure 6.34                              Figure 6.35

Display the part in its formed state. Drag the **Rollback** bar to the bottom of the FeatureManager.

**Save** the BASE.

### 6.8.3    Create the Right Tab Wall

The BASE requires two long tab walls. Create the right tab wall as an *Extruded-Thin* feature.

Drag the **Rollback** bar upward before *Sheet-Metal1* and after *Rip1*, Figure 6.36a.

Select the *Sketch* plane. Click the **outside right face**, Figure 6.36b.

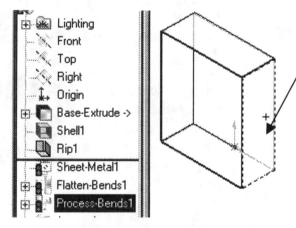

Figure 6.36a                    Figure 6.36b

Create the *Sketch*. Click the **Sketch** icon. Click the **front vertical right edge**, Figure 6.37. Click the **Convert Entities** icon.

Figure 6.37

Extrude the *Sketch*. Click the **Extrude** icon. Click the **Thin Feature** tab. Click the **Link to Thickness** check box, Figure 6.38a.

Note: Create the *Thin* feature in the direction of an existing face, Figure 6.38b. Otherwise, the result will be a disjointed feature.

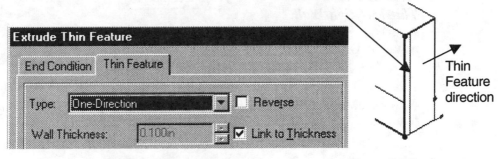

Figure 6.38a                                        Figure 6.38b

Click the **End Condition** tab. Enter **2.000** for Depth, Figure 6.39a. Display the *Extruded-Thin* feature. Click **OK** from the End Condition dialog box, Figure 6.39b.

Display the flat state. Drag the **Rollback** bar downward between the *Flatten-Bend*1 and *Process-Bend1*, Figure 6.40.

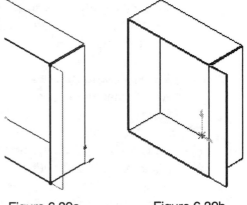

Figure 6.39a                     Figure 6.39b

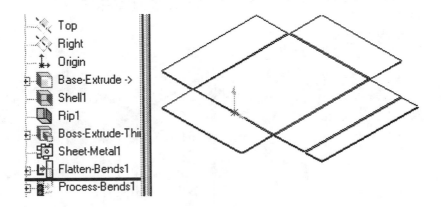

Figure 6.40

Display the *Bends*. Drag
the **Rollback** bar to the
bottom of the
FeatureManager,
Figure 6.41.

Rename *Boss-Extrude-*
*Thin1* to *RightWall*.

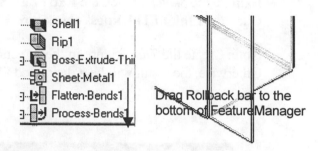

Figure 6.41

**Save** the BASE.

Note: Add additional *Thin* features after creating the *Rip* and inserting the *Sheet*
*Metal Bends*. Position the Rollback bar before *Sheet Metal1* to insert additional
features.

### 6.8.4    Create the Left Tab Wall

Create the left tab wall with the *Mirror*
*Pattern* feature. The *Mirror Pattern* feature
requires a *Mirror* plane. The *Mirror* plane is
the *Right* plane.

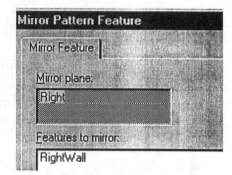

Display the *Right* plane. Right-click on the
*Right* plane in the FeatureManager. Click
**Show**.

Figure 6.42

Drag the **Rollback** bar upward before
*Sheet-Metal1* and after *RightWall*.

Create the *LeftWall*. Click the **RightWall**. Click **Insert** from the Main menu.
Click **Pattern/Mirror**. The Mirror Feature dialog box is displayed, Figure 6.42.

Select the *Mirror* plane. Click the **Right** plane, Figure 6.43a.

Display the *Mirror* feature.
Click **OK**, Figure 6.43b.

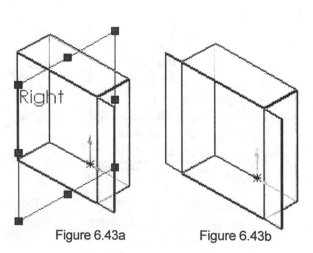

Figure 6.43a                    Figure 6.43b

Display the flat state. Drag the **Rollback** bar between *Flatten-Bends1* and *Process-Bends*1 Figure 6.44.

Rename *Mirror1* to *LeftWall*.

Drag the **Rollback** bar to the bottom of the FeatureManager. Click **Rebuild**.

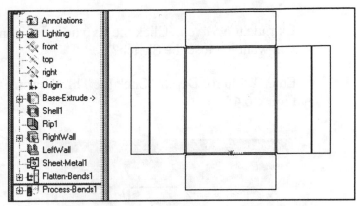

Figure 6.44

**Save** the BASE.

### 6.8.5  Create the Short Tab

Short tabs are commonly used in sheet metal parts. Create tabs by attaching a small piece of material to an existing wall. Tabs may contain holes, cuts and bends.

Create a tab. Insert the tab before the *SheetMetal Bend* feature. Drag the **Rollback** bar upward between *Sheet-Metal1* and the *LeftWall*.

Create an *Extruded-Thin* feature. Select the *Sketch* plane. Click the **inside bottom face**, Figure 6.45.

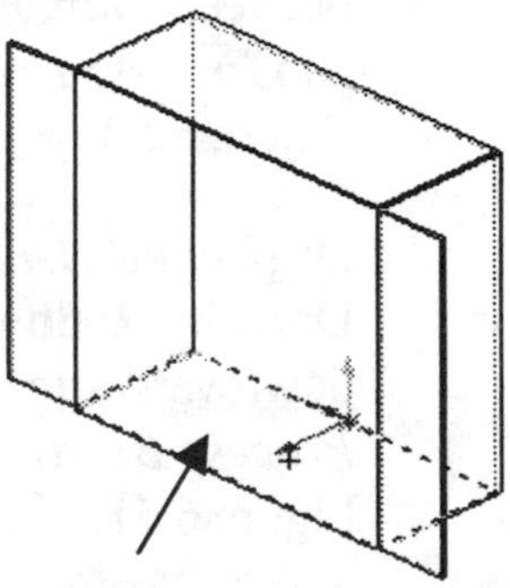

Create the *Sketch*. Click the **Sketch** icon. Click the **front bottom edge**, Figure 6.46. Click the **Convert Entities** icon. Drag both line **endpoints** towards the center.

Figure 6.45

Add a dimension. Click the **Dimension** icon. Enter **2.000**.

Create a point. Click the **Point** icon. Click the intersection of the sketched **line** and the *Right* plane, Figure 6.47.

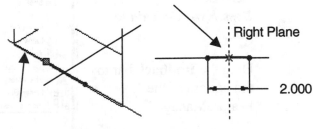

Figure 6.46                    Figure 6.47

Add geometric relations. Click the **Add Relations** icon. Create a coincident relation between the **Point** and the *Right* plane. Add a mid point relation between the Point and the sketched line. Click the **Point**. Click the **sketched line**.

Extrude the *Sketch*. Click the **Extrude** icon. Click **No** to the question, "Do you want to close the Sketch?".

Enter **1.000** for Depth. Click the **Thin Feature** tab. Click the **Reverse** check box, Figure 6.48.

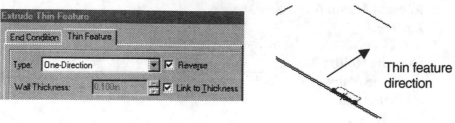

Figure 6.48

Display the *Extruded-Thin* feature. Click **OK** from the End Condition dialog box, Figure 6.49.

Display the BASE in the flat state. Drag the **Rollback** bar downward between *Flatten-Bend*1 and *Process-Bend1*, Figure 6.50.

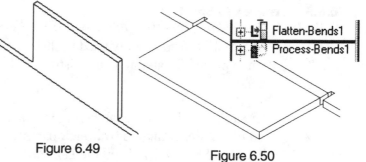

Display the *Bends*. Drag the **Rollback** bar downward after *Process-Bend1*, Figure 6.51.

Figure 6.49                    Figure 6.50

The *Bend* for the tab is created with automatic relief.

Rename **Boss-Extrude-Thin** to **Tab**.

Drag the **Rollback** bar to the bottom of the FeatureManager.

**Save** the BASE.

Click **Yes** to Rebuild the part.

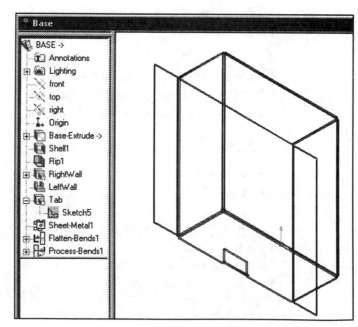

Figure 6.51

### 6.8.6   Add a Bend to an Extruded-Thin feature

Sheet metal *Bends* are added to *Extruded-Thin* features. Insert *Bends* when the part is in a flat state. *Bends* are used to create finished edges called hems.

Expand the *Process-Bends1* feature. Click the **plus** ⊞ icon to the left of *Process-Bends1*, Figure 6.52. The Flat-Sketch1 icon is displayed.

Figure 6.52

Fit the part to the Graphics window. Press the **f** key.

Insert the Bend lines, Figure 6.53. Right-click on **Flat-Sketch1**. Click **Edit Sketch**.

- Sketch the **first line** to define the *Bend* at the midpoints of the *RightWall*

- Sketch the **second line** to define the *Bend* at the midpoints of the *LeftWall*

- Sketch the **third line** to define the *Bend* at the midpoints of the *Tab*

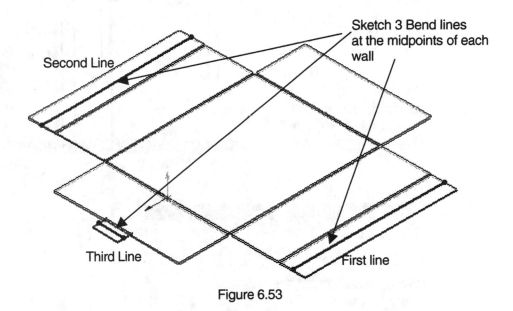

Figure 6.53

Caution: Insure that the endpoints of the created three lines are black. Verify midpoint relations.

Click the **Display Relations** icon. Review the midpoint relation for each sketched line. The *Bends* will not rebuild if the BASE size changes, Figure 6.54.

Close the *Sketch*. Click the
Sketch 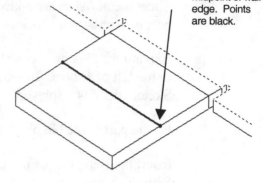 icon. A *Bend* is
inserted on each sketched line. The
*Bend* uses the default bend radius
and bend angle, Figure 6.55.

Midpoint of wall
edge. Points
are black.

*FlatBend1*, *FlatBend2* and
*FlatBend3* features are created
under the *Flat-Sketch1* feature,
Figure 6.56.

**Figure 6.54**

Modify the *RightWall* bend
direction and angle. Right-click on
**FlatBend1**. Click **Edit Definition**. The Bend from Flat dialog box is displayed,
Figure 6.57.

Enter **180** for Angle.
Click the **Bend Down**
checkbox. Display the
bend in the opposite
direction, Figure 6.58.
Click **OK**.

Note: The Bend Down
option may not be
required. The Bend
direction is toward the
back face of the BASE.

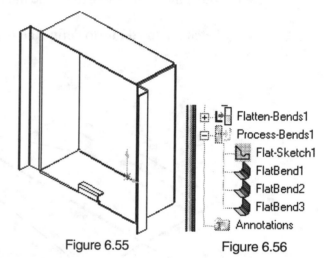

Figure 6.55                    Figure 6.56

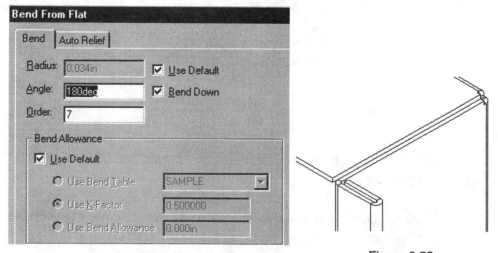

Figure 6.57                    Figure 6.58

Modify the *LeftWall* bend direction and angle. Right-click on ***FlatBend2***. Click **Edit Definition**. Enter **180** for Angle. The Bend Radius default value is 0.034. Click the **Bend Down** checkbox. Click **OK**, Figure 6.59.

**Save** the BASE.

**6.8.7    Create a Hole and a Pattern of Holes**

Sheet metal holes either are created through a punch or drill process. Each process has advantages and disadvantages:

- Cost

- Time

- Accuracy

- Etc.

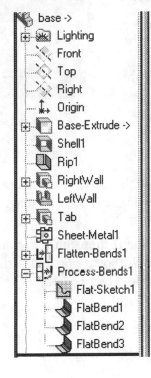

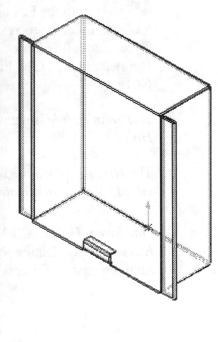

Figure 6.59

In this exercise, you investigate a *Linear Pattern* of holes. Create the *Hole* and *Linear Pattern* before the *Sheet Metal1 Bends* in order to display the *Hole* in the flat state.

Holes should be of equal size and utilize the same fasteners. Why? You need to insure a cost effect design that is price competitive. Your company must be profitable with their designs to insure financial stability and future growth.

Another important reason for fastener commonality and simplicity is the customer. The customer or service engineer does not want to supply a variety of tools for different fasteners.

A designer needs to be prepared for changes. At this time, you do not have the final overall mounting dimension locations. Design flexibility is key!

Create the hole as *Simple Hole* feature. Drag the **Rollback** bar upward before *SheetMetal1*, Figure 6.60.

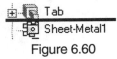

Figure 6.60

Select the *Sketch* plane. Click
the **top face**. Create the *Simple
Hole*. Click the **Simple Hole**
icon from the Features
toolbar. Enter **UptoNext** for
Type. Enter **0.250** for
Diameter. Click **OK**, Figure
6.61a.

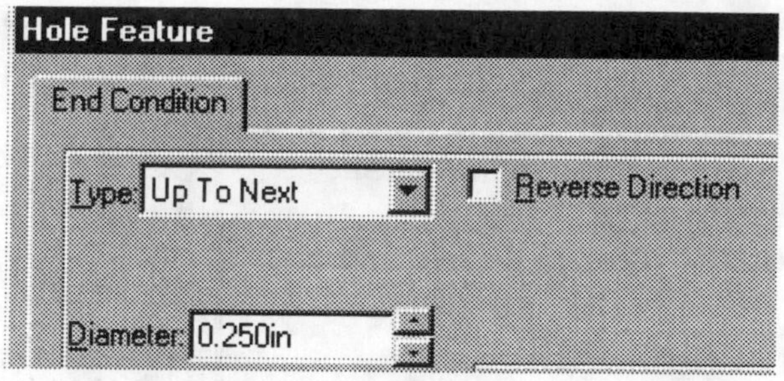

The *Simple Hole* is named
*Hole1*.

**Hole Feature**

| End Condition |

Type: Up To Next    ☐ Reverse Direction

Diameter: 0.250in

Figure 6.61a

The *Hole1* is positioned on the top face based upon the selection point. Reposition
*Hole1*. Right-click ***Hole1*** in the Feature Manager. Click **Edit Sketch**.

Dimension *Hole1*. Click the **Dimension** icon. Use the outside faces of the
*Extruded-Base* feature for the horizontal and vertical dimension references. For
each dimension, Enter **0.500**, Figure 6.61b.

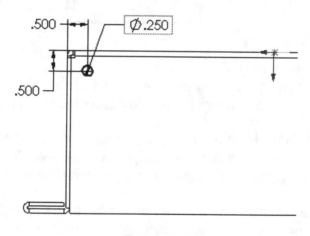

Figure 6.61b

Create a *Linear Pattern* of *Hole1*. Click ***Hole1***. Click the **Linear Pattern**
icon.

The Linear Pattern dialog box is displayed, Figure 6.62. Two pattern directions are required.

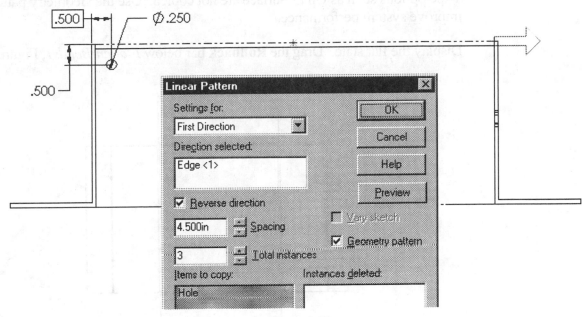

Figure 6.62

Create the pattern in the first direction. Click the **top horizontal edge** for Direction selected. Click the **Reverse direction** check box. The first direction arrow points to the left. Enter **4.500** for Spacing. Enter **3** for Total Instances.

Create the Linear Pattern in the second direction, Figure 6.63. Click the **drop down arrow** in the Setting for text box. Click **Second Direction** from the Settings list box. Click the **left edge** for Direction selected. The second direction arrow points downward. Enter **3.000** for Spacing. Enter **2** for Total Instances. Click the **Geometry pattern** check box. Click **Preview**.

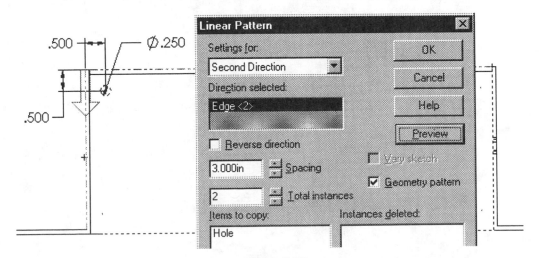

Figure 6.63

Display the *Linear Pattern*.  Click **OK.**

Note: The Geometry pattern option copies faces and edges of the seed feature. Type options such as Up to Surface are not copied.  Use the Geometry pattern to improve system performance.

Display the flat state.  Drag the **Rollback** bar below *Flatten-Bends1*, Figure 6.64.

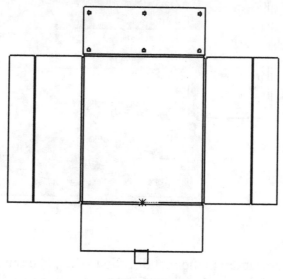

Figure 6.64

Drag the **Rollback** bar to the bottom of the FeatureManager.

Suppress *Hole1*.  Right-click ***Hole1*** in the FeatureManager.  Click **Properties**. Click **Suppressed**.  Click **OK**.  The *Linear Pattern* is suppressed.  The *Linear Pattern* depends upon *Hole1*.

**Save** the BASE.

### 6.8.8   Add a Die Cutout Palette Feature

The Palette Feature directory contains examples of predefined sheet metal shapes.  Create a die cut on the *RightWall* of the BASE.  This is for a data cable.  The team will discuss sealing issues at a later date.

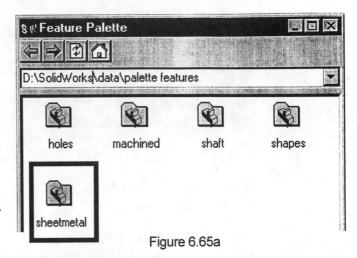

Figure 6.65a

Display the *Isometric* view.
Click the **Isometric** icon.

Click **Tools** from the Main menu.  Click **Feature Palette**.

Click the **PaletteFeatures** folder.  Click the **Sheetmetal** folder, Figure 6.65a.

The Feature Palette Sheetmetal directory is displayed, Figure 6.65b.

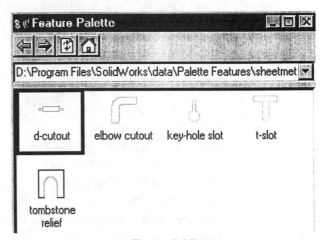

Figure 6.65b

Drag the **d-cutout** to the Graphics window.  Release the left mouse button on the **right outside face** of the BASE, Figure 6.66.

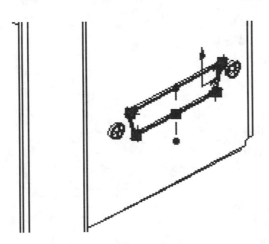

Figure 6.66

Display the *Right* view.  Click the **Right** icon.  Display the *Front* and *Top* reference planes.  Right-click **Show**.

The Edit This Sketch dialog box is displayed, Figure 6.67.

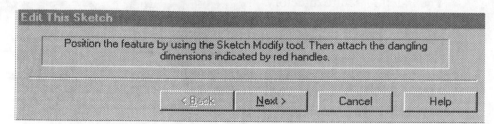

Figure 6.67

First position the *Sketch*, then dimension the *Sketch*.

Rotate the d-cutout. Click **Tools** from the Main menu. Click **Sketch Tools**, **Modify**. The Modify Sketch dialog box is displayed, Figure 6.68.

The mouse pointer displays the

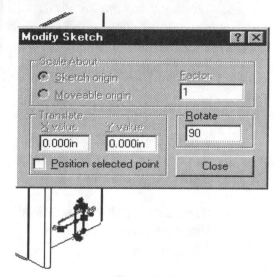

modify move/rotate icon Click and drag the **Right mouse button** to rotate the d-cutout around its black 90 degrees axis. Click the **Left mouse button** to move the d-cutout upward. Click **Close**.

Figure 6.68

Dimension the d-cutout. Click the **Dimension** icon. Create a horizontal dimension. Click the **mid point** of the d-cutout. Click the *Front* plane. Enter **1.000**. Click the **mid point** of the d-cutout. Click the *Top* plane. Enter **2.000**, Figure 6.69. Click **Next** from the Edit This Sketch dialog box.

Dimension center point to the Top and Front reference planes

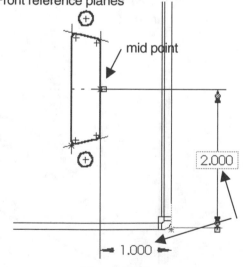

Figure 6.69

Note: With thin sheet metal parts, select the dimension references with care. Use the Zoom and Rotate commands to view the correct edge. Use reference planes to create dimensions. The planes do not change during the flat and formed states.

Accept the default dimension for the d-cutout, Figure 6.70.  Click **Apply**.
Display the d-cutout, Figure 6.71.  Click **Finish**.

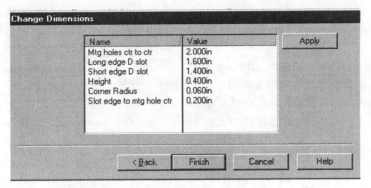

| Name | Value |
|------|-------|
| Mtg holes ctr to ctr | 2.000in |
| Long edge D slot | 1.600in |
| Short edge D slot | 1.400in |
| Height | 0.400in |
| Corner Radius | 0.060in |
| Slot edge to mtg hole ctr | 0.200in |

Figure 6.70

The d-cutout goes through the *RightWall* and
*LeftWall*.

Through All is the current Type option.
Change the d-cutout to only go through the
*RightWall*.

Right-click on the **d-cutout** in the
FeatureManager.  Click **Edit Definition**.  Click
**UptoNext** from the Type drop down list.  Click
**OK**.

The d-cutout1 is not displayed in the flat state.
The d-cutout1 was created after the
*SheetMetal1*.

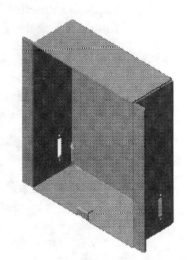

Figure 6.71

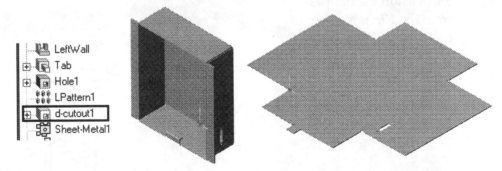

Figure 6.72a                                         Figure 6.72b

Drag **d-cutout1** upward in the FeatureManager.  Position **d-cutout1** before the
*Sheet Metal1 Bend*s, Figure 6.71a.  Display the flat state.  Drag the **Rollback** bar
downward between *Flatten Bends* and *Process Bends*, Figure 6.72b.

It is cost effective to perform the required cut in the flat state.

### 6.8.9　Add a Dimple Formed Feature

The Palette forming tool directory contains numerous sheet metal forming shapes. In SolidWorks, the forming tools are inserted after the *Bends* are processed.

Create a formed **dimple** on the top of the BASE. Click **Tools** from the Main menu. Click **Feature Palette**. Click the **PaletteFormingTools** folder. Click the **Embosses** folder.

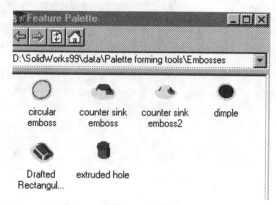

Figure 6.73

The Embosses directory is displayed, Figure 6.73. Drag the **dimple** to the top face of the BASE, Figure 6.74a.

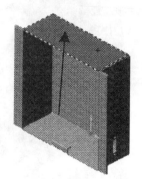

Figure 6.74a

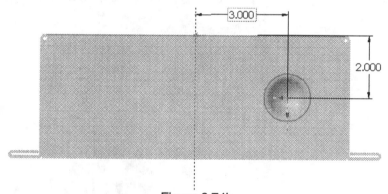

Figure 6.74b

**Dimension** the dimple from the *Right* plane of the BASE. Enter **3.000**.

Dimension the dimple from the *Front* plane of the BASE. Enter **2.000**, Figure 6.74b.

Display the dimple, Figure 6.75. Click **Finish**.

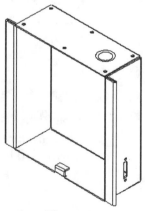

Figure 6.75

Double-click on the **dimple**. Change the dimple diameter dimension. Enter **0.500**. Change the depth dimension. Enter **0.130**, Figure 6.76. Click **Rebuild**.

The BASE is complete.

**Save** the BASE.

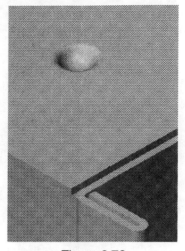

Figure 6.76

## 6.9  Manufacturing Considerations

How do you determine the size and shape of the dimple form? Are additional die cuts or forms required in the project?

Work with a sheet metal manufacturer. Ask questions. What are the standards? What tooling is in stock? Inquire on form advantages and disadvantages.

One company that has taken design and form information to the Internet is Lehi Sheetmetal, Westboro, MA. (www.lehisheetmetal.com). Standard dies, punches and manufacturing equipment: such as brakes and turrets are listed in the Engineering helpers section. Dimensions are given for their standard forms.

The form tool for this project creates a 0.500" x 0.130" dimple. The tool is commercially available.

If a custom form is required, most custom sheet metal manufacturers can accommodate your requirement. Example: Wilson Tool, Great Bear Lake, MN (www.wilsontool.com). Note: However, you will be charged for the tool.

How do you select material? Consider the following:

- Strength

- Fit

- Bend Characteristics

- Weight

- Cost

All of these factors influence material selection. Raw aluminum is a commodity. Large manufacturers such as Alcoa and Reynolds sell to a material supplier, such as *Pierce Aluminum. Pierce Aluminum in turn, sells material of different sizes and shapes to distributors and other manufacturers. Material is sold in sheets or cut to size in rolls, Figure 6.77.

| **ALUMINUM SHEET**<br>**NON HEAT TREATABLE, 1100-0**<br>**QQ-A-250/1  ASTM B 209** | |
|---|---|
| All thickness and widths available from coil for custom blanks | |
| SIZE IN INCHES | WGT/SHEET |
| .032 x 36 x 96 | 11.05 |
| .040 x 36 x 96 | 13.82 |
| .040 x 48 x 144 | 27.65 |
| .050 x 36 x 96 | 17.28 |
| .050 x 48 x 144 | 34.56 |
| .063 x 36 x 96 | 21.77 |
| .063 x 48 x 144 | 43.55 |
| .080 x 36 x 96 | 27.65 |
| .080 x 48 x 144 | 55.30 |
| .090 x 36 x 144 | 46.66 |
| .090 x 48 x 144 | 62.26 |
| .100 x 36 x 96 | 34.56 |
| .125 x 48 x 144 | 86.40 |
| .125 x 60 x 144 | 108.00 |
| .190 x 36 x 144 | 131.33 |

Figure 6.77
*Courtesy of Piece Aluminum Co, Inc.
Canton, MA

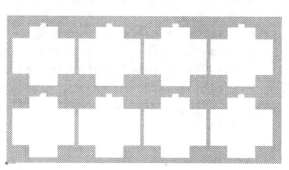

Figure 6.78a

Figure 6.78b

Sheet metal manufacturers work with standard 8'stock sheets. For larger quantities, the material is usually supplied in rolls. For a few cents more per pound, sheet metal manufacturers request the supplier to shear the material to a custom size.

The overall size for the BASE is approximately 20.7" x 21.7". The overall flat box width and height is less than 48"/2"= 24". Eight units are accommodate by one 4' x 8' sheet, Figure 6.78a.

Do not waste raw material. Optimize for sheet manufacturing. Example: Flat box height 25". Three to four units are accommodated by one 4' x 8' sheet, Figure 6.78b.

In Project 1, you were required to be cognizant of the manufacturing process for machined parts. In Project 4 and 5 you created and purchased parts. Whether a sheet metal part is produced in or out of house, knowledge of the materials, forms and layout provides the best cost effective design to the customer.

You require both a protective and cosmetic finish for the Aluminum BOX. Review options with the sheet metal vendor. Parts are anodized. Black and clear anodized finishes are the most common. In harsh environments, parts are covered with a protective coating such as Teflon® or Halon®.

The finish adds thickness to the material. A few thousandths could cause problems in the assembly. Think about the finish before the design begins.

### 6.10 Create the COVER

The size of the COVER is dependent on the size of the BASE. The parent component is the BOX. Create the COVER in context of the BOX. Extract edges from the BASE. Create the *Base-Extruded-Thin* feature for the COVER on the *Right* plane. The depth of the COVER references the corners of the BASE. Insert sheet metal *Bends* to complete the part.

### 6.10.1 Update the BASE Component in the BOX Assembly

Return to the assembly. Click **Window** from the Main menu.

Click **BASE–in-BOX** assembly from the file list. Click **YES** to the question, Would you like to rebuild the assembly now?", Figure 6.79a.

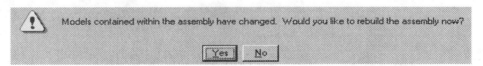

Figure 6.79a

Update the BASE component in the assembly, Figure 6.79b.

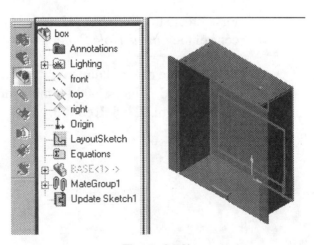

Figure 6.79b

### 6.10.2 Verify the BASE

Verify the BASE component for the large BOX.

Remove the suppress state from the *Hole1* and *Linear Pattern*. Right-click on ***Linear Pattern***. Uncheck the **Suppress** check box.

Display all dimensions. Right-click the **Annotations** folder in the BOX FeatureManager. Click **Show Feature Dimensions**, Figure 6.80a.

Modify the motherboard width and height dimensions. Double-click on the mb-width, **8.000**. Enter **20.000**. Double-click on the mb-height, **6.000**. Enter **24.000**.

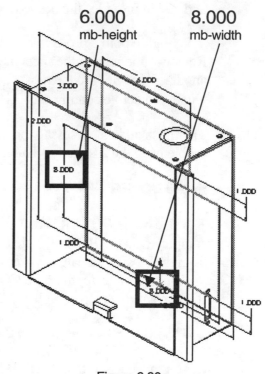

6.000          8.000
mb-height      mb-width

Figure 6.80a

Click **Rebuild**, Figure 6.80b.

The holes are not equally spaced. The *Layout Sketch* controls the overall dimensions of the BOX.

Control the *Linear Hole Pattern* with the *Equation* feature.

The COVER contains the same *Linear Hole Pattern* as the BOX.

First, create the COVER. Then add the *Equations* to control the *Linear Hole Pattern*.

Return to the original dimensions for the motherboard. Enter **8.000** for mb-width. Enter **6.000** for mb-height. Click **Rebuild**.

**Save** the BOX.

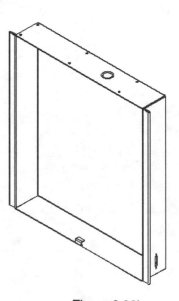

Figure 6.80b

If rebuild errors occur, review the following:

- For flat bends, the bend line endpoints must intersect the edges of the wall, Figure 6.80c.

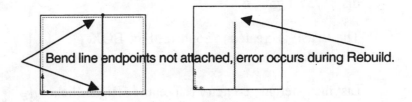

Bend line endpoints not attached, error occurs during Rebuild.

Figure 6.80c

- The Bend radius is too small, increase the Bend Radius, Figure 6.80d.

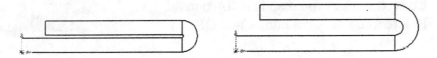

Figure 6.80d

- The part intersects itself, Figure 6.80e.

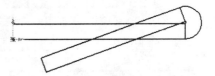

Figure 6.80e

### 6.10.3 The Layout Sketch Position

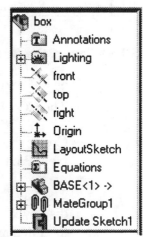

Figure 6.81a

The *Layout Sketch* contains dimensions and variables. The COVER references dimensions and variables.

The *Layout Sketch* is positioned between the *Origin* and *Equations*. This insures that the BOX is updated correctly, Figure 6.81a.

The *Update Entry* feature is displayed at the bottom of the FeatureManager.

The *Update Entry* contains a list of Parent/Child relationships and External geometry references. *Sketch1* is the rectangular profile extracted from the *Layout Sketch* with the Convert Entities command.

Right-click on the **Update Sketch1 in BASE**.
Click **Parent/Child**.

The Parent/Child Relationships dialog box is
displayed, Figure 6.81b.

The BASE is created in context of the BOX.
Click **Close**.

List the external geometry references. Right-click
on the **Update Sketch1 in BASE**. Click **List
External Ref**. The External Reference For:
dialog box is displayed, Figure 6.81c. The lines
extracted from the feature *Layout Sketch* using the
Convert Entities command are listed in the
Referenced Entity column. Click **OK**.

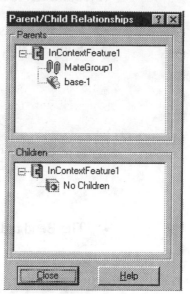

Figure 6.81b

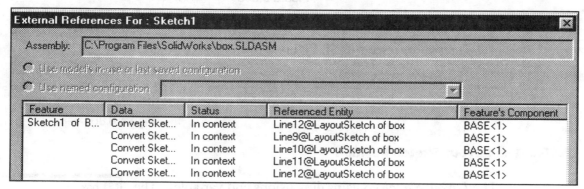

Figure 6.81c

### 6.10.4  Insert the Second Component

The COVER is the second component. Create the COVER inside the assembly.
The COVER references the BASE.

Create the COVER. Click **Insert** from the Main menu. Click **Component**. Click
**New Part**.

The Save As dialog box is
displayed. Enter **COVER** in
the Filename text box,
Figure 6.82. Click **Save**.

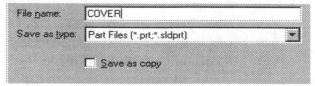

Figure 6.82

Select the In Place Mate plane. Click the ***Right*** plane of the BOX in the FeatureManager, Figure 6.83. The *Right* plane of the component is mated with the *Right* plane of the BOX assembly.

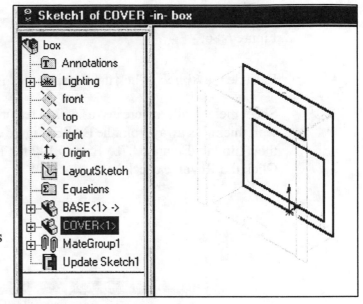

The COVER is added to the FeatureManager.

The Edit Part icon is selected automatically. The text COVER appears in the FeatureManager. The text is displayed in light gray. This indicates that the part is actively being edited.

Figure 6.83

The current *Sketch* plane is the *Right* plane. The current *Sketch* name is *Sketch1*. This is indicated on the current graphics window title, "Sketch1 of COVER - in-BOX." COVER is the name of the component created in the context of the BOX assembly.

The Sketch icon is automatically selected.

Hide the *Layout Sketch*. Right-click on the **Layout Sketch**. Click **Hide**. View the edges of the BASE. Click **Hidden Line Removed**.

Create the *Extruded-Base* feature for the COVER component.

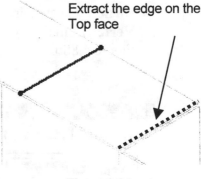

Extract the edge on the Top face

Figure 6.84

Define the profile of the COVER. Extract existing edges from the BASE. Create the *Sketch*. Click the **right top edge**, Figure 6.84. Click the **Convert Entities** icon.

View the bottom right edge. **Rotate** the assembly. Click the **right bottom edge**, Figure 6.85. Click the **Convert Entities** icon.

Figure 6.85

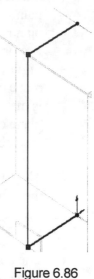

Connect the top and bottom lines.  Sketch a **vertical line**, Figure 6.86.

Extrude the *Sketch*.  Click the **Extrude** icon.

Sheet metal walls are created as a *Thin* feature.  Create the wall thickness away from the BASE, Figure 6.87a.  Click the **Thin** tab, Figure 6.87b.  Enter **0.100** for Thickness. Click the **Reverse** check box.

Figure 6.86

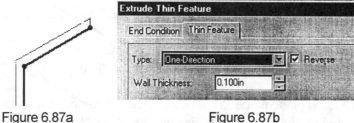

Figure 6.87a                    Figure 6.87b

The extruded depth of the COVER is dependent on the BASE.  Click the **End Condition** tab.  Click the **Both Directions** check box.  Create Direction 1.  Click **Up To Vertex** from the Type drop down list.  Click the vertex at the **top left corner**.  Create Direction 2.  Click **Direction 2**, Figure 6.88a.  Click **Up To Vertex** from the Type drop down list.  Click the vertex at the **top right corner**, Figure 6.88b.

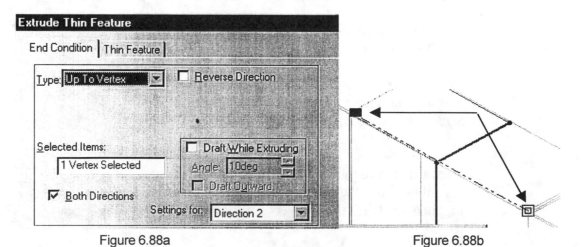

Figure 6.88a                                      Figure 6.88b

Note: The Vertex of the Cover is displayed in red.

Display the *Base-Extrude-Thin* feature, Figure 6.89a. Click **OK**.

Rename *Sketch1* in the *Base-Extrude-Thin* feature to *CoverSketch*.

Close the part. Click the **Edit Part** icon.

Click **Rebuild**.

**Save** the BOX.

The entries: *Update CoverSketch1 in cover* and *Update Base-Extrude-Thin in cover* are added to the FeatureManager, Figure 6.89b.

Figure 6.89a

Layout
Update Sketch1 in base
Update CoverSketch in cover
Update Base-Extrude-Thin in cover

Figure 6.89b

## 6.10.5 Holes

Holes are created in each part or in the context of the assembly. Example: Between two components such as the BASE and COVER.

Reference holes between multiple components in the assembly.

In this example, extract the hole geometry from the BASE *Hole1* seed feature.

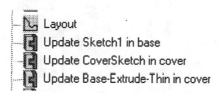

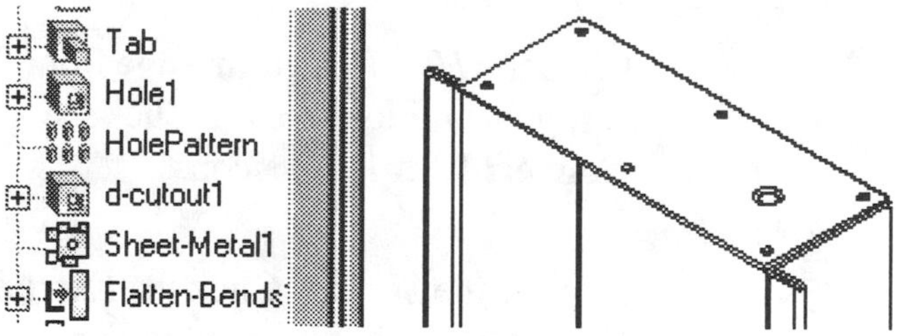

Tab
Hole1
HolePattern
d-cutout1
Sheet-Metal1
Flatten-Bends

Figure 6.90

Create a *Linear Pattern*. Use an *Equation* in the assembly to equate the pattern dimensions between the BASE and COVER. The *Equations* insure proper alignment for multiple mating components.

Expand the BASE. Click the **Plus** icon.

Rename *Linear Pattern1* to *Hole Pattern*, Figure 6.90. Click *HolePattern*. Highlight the *Hole1* seed feature. Click *Hole1*.

Names and highlighted features can be modified in the BASE while you are editing the COVER. Avoid unwanted references. Only click only the required geometry.

Review the part status in the FeatureManager. The BASE name is in black. The COVER name is in red. The Edit Part 🔲 icon indicates that you are editing the COVER, Figure 6.90a.

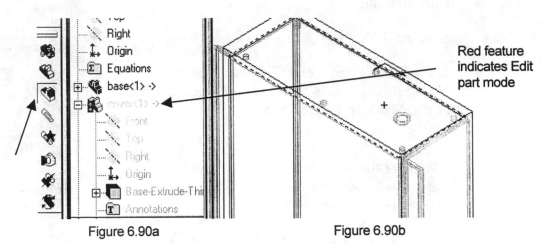

Figure 6.90a                              Figure 6.90b

**Collapse** the BASE entry in the FeatureManager.

Extract *Hole1* from the BASE.

Select the *Sketch* plane. Click the **top face of the COVER**, Figure 6.90b.

Create the Sketch. Click the **Sketch** 🔲 icon.

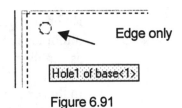

Figure 6.91

Click the *Hole1* **circular edge** from the BASE component, Figure 6.91. Click the **Convert Entities** 🔲 icon.

Extrude the *Sketch*. Click the **Extrude- Cut** 🔲 icon. Click **Through All** from the Type drop down list. Display the *Extruded-Cut*. Click **OK**, Figure 6.92. Rename *Cut-Extrude* to *Hole1*.

## 6.10.6 Insert Sheet Metal Bends

Insert *Sheet Metal Bends* from within the COVER.

Close Edit Part. Click the **Edit Part** 🔲 icon. The COVER and its features are displayed in black.

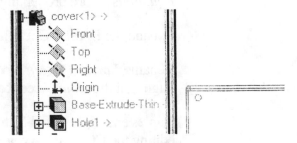

Figure 6.92

Right-click the **COVER** from the FeatureManager. Click **Open COVER.sldprt**.

Display the *Isometric* view.

Insert the *Bends*. Click the **bottom face**, Figure 6.93. Click the **Insert Bend** icon from the Feature toolbar. Accept the default values. Click **OK**. Drag the **Rollback** bar between *Flatten-Bends1* and *Process-Bends2*, Figure 6.94.

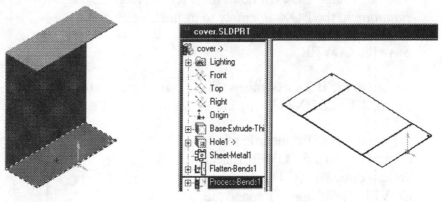

Figure 6.93                    Figure 6.94

### 6.10.7 Create a Linear Pattern of Holes

Create a *Linear Pattern* of holes. Drag the **Rollback** bar upward between *Hole1* and *Sheet-Metal1*.

Use the dimension values created for the BASE *Linear Pattern*, Figure 6.95a.

Create the Linear Pattern. Click the **Linear Pattern** icon. Create Direction 1. Enter **4.500** for Spacing. Enter **3** for Total Instances. Create Direction 2. Enter **3.000** for Spacing. Enter **2** for Total Instances.

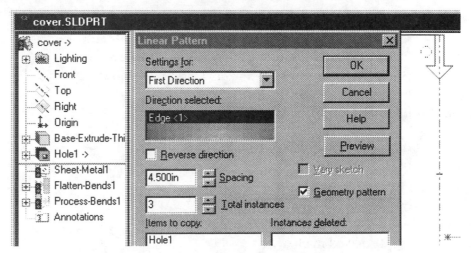

Figure 6.95a

Display the *Linear Pattern*, Figure 6.95b.  Click **OK**.

Rename *LinearPattern* to *HolePattern*.

### 6.11  Use Equations in the Assembly

The BASE and COVER holes must remain aligned.  Use an *Equation* in the BOX assembly to insure hole alignment.

**Save** the COVER.

Return to the BOX assembly.  Click **Window** from the Main menu.  Click **BOX**.  Update the assembly.  Click **Yes**.

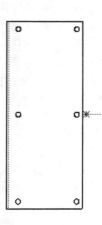

Figure 6.95b

Equate the BASE Linear Pattern dimensions to the COVER Linear Pattern dimensions.  Display all COVER dimensions.  Expand the COVER.  Click the **Plus** ⊞ icon. Right-click on the **Annotations** folder.

Click **Show Feature Dimension**.  The Linear Pattern COVER dimensions are in the same location as the Linear Pattern Dimensions of the BASE. Drag the **4.500** and **3.000** dimension upward, Figure 6.96a.

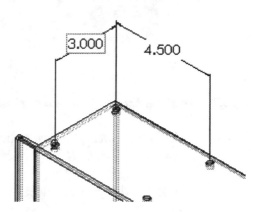

Figure 6.96a

Display the dimensions for the BASE. Click **Show Feature Dimension**. Click on the *Linear HolePattern* feature, Figure 6.96b.

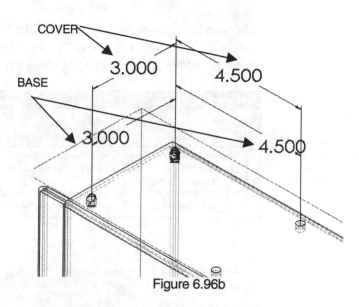

Figure 6.96b

Be consistent. The COVER dimensions are dependent on the BASE dimensions.

*Equation* 1 and *Equation* 2 were created in the *Layout Sketch*.

Create the third *Equation*.
Right-click the *Equations* folder in the FeatureManager. Click **Add Equations**.

Click the COVER dimension. Click **4.500**. **Enter** = from the keypad. Click the BASE dimension. Click **4.500**. Click **OK**.

Create the forth *Equation*. Click the COVER dimension. Click **3.000**. Click **Add** button. Click the BASE dimension. **Enter** = from the keypad. Click **3.000**. Click **OK**.

"D3@HolePattern@cover.Part"="D3@HolePattern@base.Part"
"D4@HolePattern@cover.Part"="D4@HolePattern@base.Part"

Figure 6.97

View the two new *Equations*, Figure 6.97. The two new *Equations* insure that the Pattern dimensions remain equal for the COVER and BASE.

When designing, think ahead! What happens if the size of the BOX changes? Your customer requested three difference BOX sizes. How do you insure that the *HolePattern* remains equally spaced for all sizes? Answer: Create an additional *Equation*.

Note: The first edge direction of the *HolePattern* was selected along the width of the BOX. The hole center point is 0.500".

Create the fifth *Equation*.

Enter **"D3@HolePattern@base.Part"= "box-width@LayoutSketch"/2 - 0.500**

Click **OK**.

*Equation 5* is in the wrong location. Position *Equation 3* before *Equation 5*. Why? The *HolePattern* for the COVER and the BASE must be the same.

The windows short cut keys Ctrl X/Ctrl V are used to cut/paste *Equations 5*.
Select *Equation 5*. Click **Ctrl X**. Position the **mouse pointer** before *Equation 3*.
Paste Equation 3. Click **Ctrl V**. Figure 6.98.

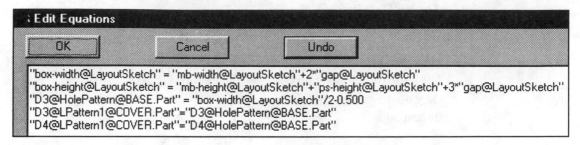

**Edit Equations**

OK          Cancel          Undo

```
"box-width@LayoutSketch" = "mb-width@LayoutSketch"+2*"gap@LayoutSketch"
"box-height@LayoutSketch" = "mb-height@LayoutSketch"+"ps-height@LayoutSketch"+3*"gap@LayoutSketch"
"D3@HolePattern@BASE.Part" = "box-width@LayoutSketch"/2-0.500
"D3@LPattern1@COVER.Part"="D3@HolePattern@BASE.Part"
"D4@LPattern1@COVER.Part"="D4@HolePattern@BASE.Part"
```

Figure 6.98

Return to the
Equations dialog
box, Figure 6.99.
Click **OK**.

What feature in
the BOX
controls the
location of the
holes? Answer:
The *Layout
Sketch*.

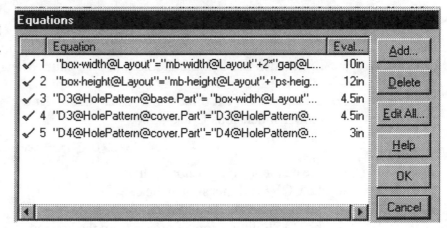

**Equations**

| | Equation | Eval... |
|---|---|---|
| ✓ 1 | "box-width@Layout"="mb-width@Layout"+2*"gap@L... | 10in |
| ✓ 2 | "box-height@Layout"="mb-height@Layout"+"ps-heig... | 12in |
| ✓ 3 | "D3@HolePattern@base.Part"= "box-width@Layout"... | 4.5in |
| ✓ 4 | "D3@HolePattern@cover.Part"="D3@HolePattern@... | 4.5in |
| ✓ 5 | "D4@HolePattern@cover.Part"="D4@HolePattern@... | 3in |

Add... Delete Edit All... Help OK Cancel

Figure 6.99

Verify the
*HolePattern* location for the BASE and COVER, Figure 6.100.

Double-click the ***Layout*** *Sketch*. Click the mb-width dimension. Click **8.000**.
Enter **16.000**. Click **Rebuild**.

Click the mb-height. Click **16.000**. Enter **24.000**. Click **Rebuild**.

Return to the original mother board width. Enter **8.000**. Click **Rebuild**.

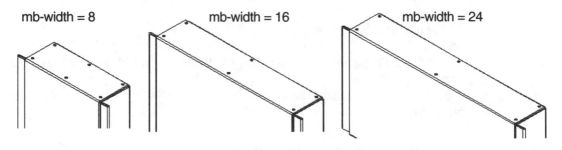

mb-width = 8      mb-width = 16      mb-width = 24

Figure 6.100

## 6.12 Add an Assembly Hole Feature

Additional holes and cuts are created after the forming process.

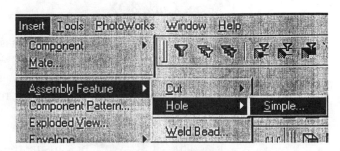

Figure 6.101

A through *Hole* is required from the front of the COVER to the back of the BASE. The *Hole* must go through the BASE *Tab*.

Create a *Simple Assembly Hole* feature. Click the **front face** of the COVER. Click **Insert** from the Main menu. Click **Assembly Feature**. Click **Hole**. Click **Simple**, Figure 6.101.

Click **Through All** from the Type drop down list. Enter diamctcr. Enter **0.250**. Click **OK**.

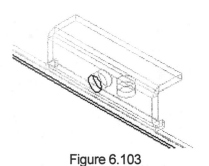

Figure 6.102

The *Simple Assembly Hole* feature, *Hole1* is added to the BOX assembly.

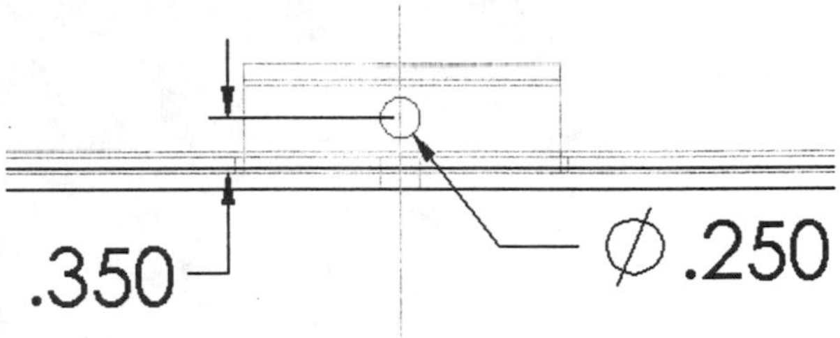

Position *Hole1*. Right-click on *Hole1* in the FeatureManager. Click **Edit Sketch**. Add a vertical **dimension** between the center point and the *Origin*, Figure 6.102. Add a coincident relation between the **center point** of *Hole1* and the ***Right*** plane.

Display *Hole1*, Figure 6.103. Click the **Sketch**   icon.

**Save** the BOX.

Figure 6.103

## 6.13 Add Components

SCREWs are required to complete the assembly. Reference the SCREW to the *HolePattern* of the COVER. Complete the assembly pattern. Mate the SCREW to the *Hole1* seed feature.

In Project 4, exercise 2 you created a simple SCREW, Figure 6.104.

Figure 6.104

Display the BOX.  Click **Window** from the Main
menu.  Click **BOX**.

Insert the SCREW into the BOX assembly.

Click **Insert** from the Main menu.  Click **Component**.
Click **From File**.  Enter the part name **SCREW** in the
File Open dialog box.

Place the SCREW.  Click inside the Graphics window
next to the **back left corner** of the BOX,
Figure 6.105.

Figure 6.105

Mate the SCREW to the COVER.  Add a Concentric Mate.  Click the **cylindrical
face** of the SCREW.  Click the **cylindrical face** of *HOLE1* of the COVER,
Figure 6.106a.

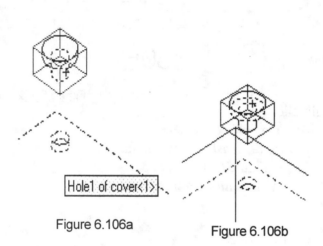

Caution: Select the face of the
*Hole1* COVER.  Do not select
the *Hole1* BASE.

Add a Coincident Mate.  Click
the **bottom head face** of the
SCREW.  Click the **top face** of
the COVER, Figure 6.106b.

Add a Parallel Mate.  Click the
*Right* plane of the SCREW.
Click the *Right* plane of the
BOX assembly.  The SCREW
is fully constrained,
Figure 6.107.

Figure 6.106a          Figure 6.106b

**Save** the BOX.

### 6.14  Add a Component Pattern in the Assembly

Multiple copies of a component in an assembly
are defined as a *Component Pattern*.  *A
Component Pattern* is created from an existing
part feature *Pattern*.  This pattern is called a
*Derived Pattern*.  The component COVER
contains the *Linear Pattern*.  There are 6 holes.  The component SCREW
references the *Hole1* seed feature.  The *Component Pattern* displays 6 SCREWS.

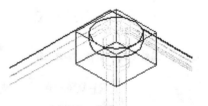

Figure 6.107

Create a *Derived Pattern*.
Click **Insert** from the Main
menu. Click **Component
Pattern**. The Pattern Type
dialog box is displayed,
Figure 6.108. Click the **Use
an existing feature pattern
(Derived)** check box. Click
**Next**.

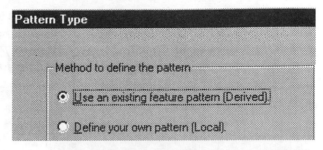

Figure 6.108

The Derived Component Pattern
dialog box is displayed,
Figure 6.109. Click the **SCREW**
for the Seed Component text box.
Click the *HolePattern* of the
COVER from the Pattern Feature
text box.

Note: Insure that the *HolePattern* is
from the COVER component and
not from the BASE component.

Display the *Derived LPattern*,
Figure 6.110. Click **Finish**.

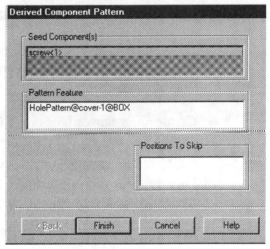

Figure 6.109

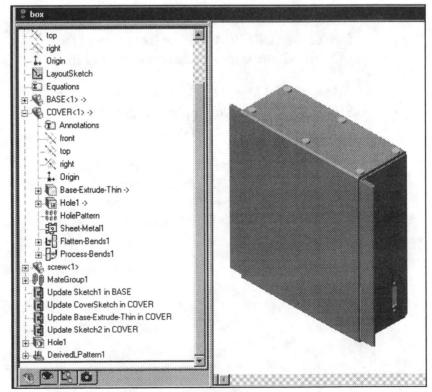

Figure 6.110

The additional SCREW Instances are located under the *Derived LPattern1* in the Feature Manager. Display the Instances. Expand *Derived LPattern1*, Figure 6.111. Click the **Plus** ⊞ icon to the left of *Derived LPattern1* in the FeatureManager.

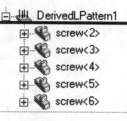

Figure 6.111

**Save** the BOX.

## 6.15  Create a second SCREW Instance

A second SCREW Instance is required to complete the project. The SCREW secures the COVER to the *Tab* of the BASE.

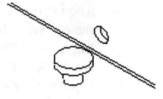

Figure 6.112

Create an instance. Click the **SCREW<1>** ⊞ 🔩 screw<1> icon. Hold the **Ctrl** key down. Drag SCREW<1> into the Graphics window, Figure 6.112. Release the **Ctrl** key. Release the **left mouse button**.

Mate the SCREW to the COVER. Add a Concentric Mate. Click the **cylindrical face** of the SCREW. Click the **cylindrical face** of *HOLE1*, Figure 6.113a.

Add a Distance Mate. Click the **bottom head face** of the SCREW. Click the **front face** of the COVER. Enter **0** for Distance. Figure 6.113b.

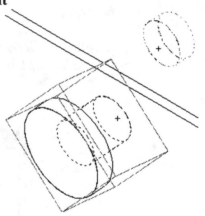

Add a Parallel Mate. Click the *Right* plane of the SCREW. Click the *Right* plane of the BOX assembly. SCREW<7> is fully constrained.

Figure 6.113a

Note: Assembly reference planes are used to create mates. The *Right* reference plane of the BOX is used to create a parallel mate between SCREW<7> and the assembly *Hole* feature.

**Save** the BOX.

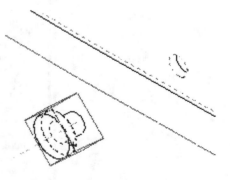

Figure 6.113b

A second MateGroup entry is created in the FeatureManager. **Expand** MateGroup2, Figure 6.114. MateGroup2 contains the SCREW<7> Mates.

Part order creation is important. The *Layout Sketch* must occur before the BASE. The first in context feature references the BASE. The remaining in context features reference the COVER.

The *Hole1* assembly feature occurs after the BASE and COVER.

In this example, the system solves the relationships in MateGroup1 with the BASE, COVER and SCREW<1>. Then the system solves the relationships in MateGroup2 with the COVER and SCREW<7>.

```
screw<1>
screw<7>
MateGroup1
    Inplace1 (BASE<1>,front)
    Inplace2 (COVER<1>,right)
    Concentric1 (COVER<1>,screw<1>)
    Coincident7 (COVER<1>,screw<1>)
    Parallel1 (COVER<1>,screw<1>)
    Update Sketch1 in BASE
    Update CoverSketch in COVER
    Update Base-Extrude-Thin in COVER
    Update Sketch2 in COVER
    Hole1
    DerivedLPattern1
MateGroup2
    Concentric2 (COVER<1>,screw<7>)
    Parallel2 (screw<7>,right)
    Distance1 (COVER<1>,screw<7>)
```

Figure 6.114

## 6.16 Design Table

The first BOX is complete! Now create a family of parts. A family of parts represents different configurations of a part or assembly. Use the Design Table to create the part family. Each configuration is called an Instance. You can create many configurations of a part or assembly by applying various dimension values.

Creating a Design Table requires Microsoft EXCEL 97 or a later version.

Create a Design Table. Click **Tools** from the Main menu. Click **Options**. Click the **General** tab. Verify that the Edit Design Tables in a separate Window check box is not selected, Figure 6.115.

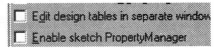

Figure 6.115

Display the required dimensions.
Right-click on the BOX **Annotations**
folder in the FeatureManager. Click
**Show Feature Dimensions**,
Figure 6.116.

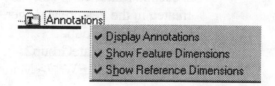

Figure 6.116

The *Layout Sketch* dimensions are
displayed, Figure 6.117a.

Display the BOX in an *Isometric*
view. **Move** the BOX to the lower
half of the Graphics window. Drag
the **dimensions** outward for
improved visibility.

Note: When creating the Design
Table, select the dimensions. Do
not select the lines.

Create a new Design Table. Click
**Insert** from the Main menu. Click
**New Design Table**.

An EXCEL worksheet appears in
the left-hand corner of the
Graphics window, Figure 6.117b.

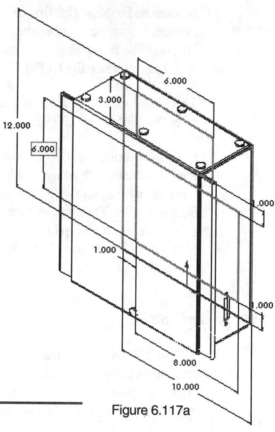

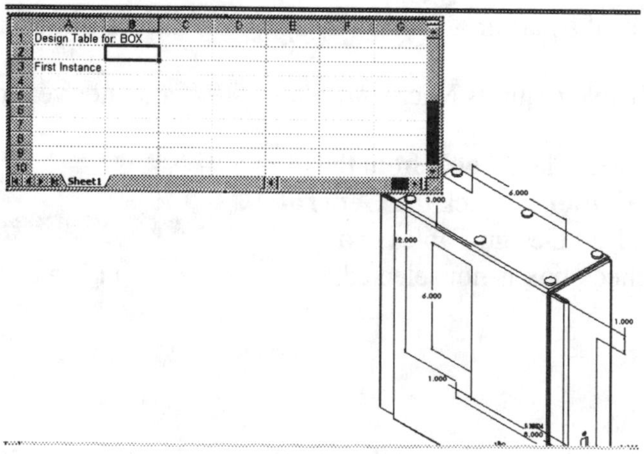

Figure 6.117a

Figure 6.117b

Cell A3 contains the text: First Instance.  Cell B2 is the active cell, Figure 6.118a.

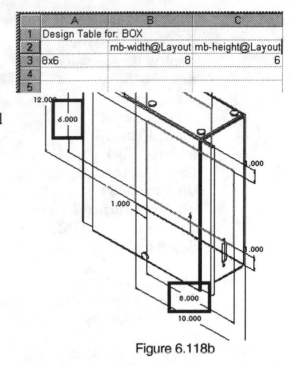

|   | A | B | C | D | E |
|---|---|---|---|---|---|
| 1 | Design Table for: BOX | | | | |
| 2 | | | | | |
| 3 | First Instance | | | | |
| 4 | | | | | |

Figure 6.118a

Note: When you click the dimension in the Graphics window, the variable name is displayed in Row 2.  Enter values for the First Instance, Figure 6.118b. Click **mb-width dimension**.  The variable name, mb-width@Layout is added to cell B2.  The value 8 is added to cell B3.

Position the mouse pointer at cell C3. Press the **Tab** key.  Click the **mb-height dimension**.  The variable name, mb-height@Layout is added to cell C2.

Value 6 is added to cell C3.

Click cell A3 text: **First Instance**. Enter **8 x 6**.  The first Instance is complete.

|   | A | B | C |
|---|---|---|---|
| 1 | Design Table for: BOX | | |
| 2 | | mb-width@Layout | mb-height@Layout |
| 3 | 8x6 | 8 | 6 |
| 4 | | | |
| 5 | | | |

Figure 6.118b

Complete the Design Table, Figure 6.119.  Click cell **A4**.  Enter **16 x 20**.

Click cell **B4**.  Enter **16**.  Click cell **C4**.  Enter **20**.

Click cell **A5**.  Enter **20x28**.  Click cell **B5**.  Enter **20**.  Click cell **C5**.  Enter **28**.

Close the Design Table.  Click inside the **Graphics window**.

|   | A | B | C |
|---|---|---|---|
| 1 | Design Table for: BOX | | |
| 2 | | mb-width@Layout | mb-height@Layout |
| 3 | 8x6 | 8 | 6 |
| 4 | 16x20 | 16 | 20 |
| 5 | 20x28 | 20 | 28 |
| 6 | | | |

Figure 6.119

Note: To edit the Design Table, click Edit from the Main menu.  Click Design Table.

Create the Instances. Click **OK** to the statement, "The design table generated the following instances", Figure 6.120a.

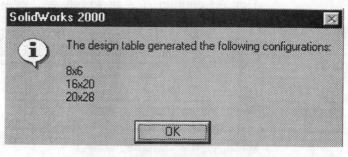

Figure 6.120a

Click the **Configuration** 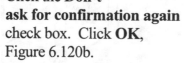 icon. The system displays the following message "Some configurations may take time to show. Click OK to show. Click Cancel to abort." Click the **Don't ask for confirmation again** check box. Click **OK**, Figure 6.120b.

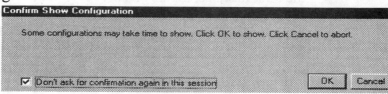

Figure 6.120b

The system displays a message that the Design Table generated the configurations. Click **OK**.

Display the configurations. Double-click the configuration **8 x 6**, Figure 6.121.

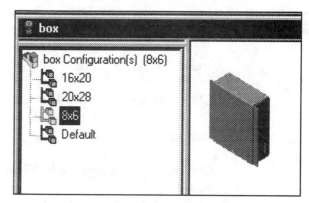

Figure 6.121

Each configuration rebuilds from values entered in the Design Table.

Display the 16 x 20 configuration.

Double-click on **16 x 20**, Figure 6.122.

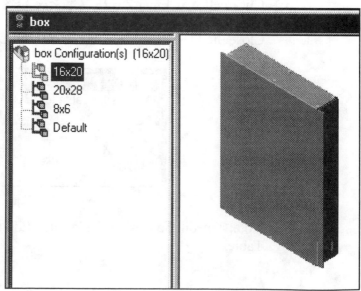

Figure 6.122

Double-click on **20 x 28**, Figure 6.123.

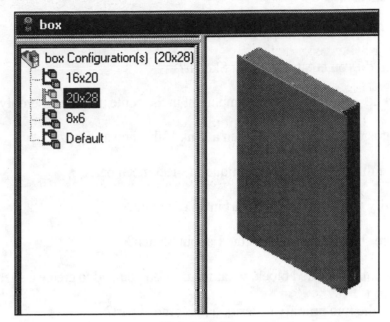

Figure 6.123

Return to the Default.  Double-click on the **Default** Configuration.

Return to the part.  Click the **Part** icon at the bottom of the FeatureManager, Figure 6.124.

Save the **BOX**.

Your design work on this project is complete.
Let's try a few examples.

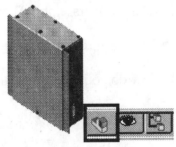

Figure 6.124

## 6.17  Questions

1.  What is a Top down approach?

2.  When do you create a Layout Sketch?

3.  How do you create a new component in the context of the assembly?

4.  What is the difference between a Link Value and an *Equation*?

5.  Name three characteristics unique to sheet metal parts.

6.  Identify the indicator for a part in an edit state?

7.  Where should you position the Layout Sketch?

8.  For a solid extruded block, what features do you add to create a sheet metal part?

9.  Name the two primary states of a sheet metal part?

10. How do you insert a formed sheet metal feature such as a dimple or louver?

11. Identify the type of information that a sheet metal manufacturer provides.

12. What is an Assembly Feature?

13. What features are required before you create the Component Pattern in the assembly?

14. How do you select an entry in the Design Table?

15. How do you display the configurations from the Design Table?


Top down assembly modeling is a complex undertaking.  Planning is key.  Before fully developing the assembly, try various configurations with a simple Layout Sketch and a few simple components.

## 6.18 Exercises

### Exercise 6.1
### Create the L-BRACKETS

Create a family of sheet metal
L-BRACKETS, Figure E6.1a.
L-BRACKETS are used in the construction
industry.

Standardize your design. Create an eight
hole L-BRACKET, Figure E6.1b. Use
*Equations* to control the hole spacing.

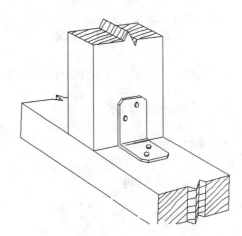

Figure E6.1a Stong-Tie
Reinforcing Brackets
"Courtesy of Simpson Strong Tie
Corporation of California"

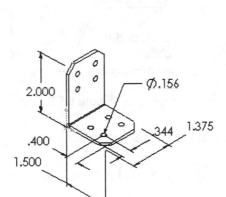

Figure E6.1b

|   | A | B | C | D | E |
|---|---|---|---|---|---|
| 1 | Design Table for: lbracket | | | | |
| 2 |  | height@Sketch1 | width@Sketch1 | depth@Base-Extrude-Thin | hole1dia@Sketch2 |
| 3 | First Instance | 2 | 1.5 | 1.375 | 0.156 |
| 4 | Small2x1.5x1 | 2 | 1.5 | 1 | 0.156 |
| 5 | Medium3x2x2 | 3 | 2 | 2 | 0.156 |
| 6 | Large4x3x4 | 4 | 3 | 4 | 0.25 |

L-BRACKET FAMILY TABLE

### Exercise 6.2
### Create a Layout Sketch

Create a *Layout Sketch* for a Top down design in a new appliance or consumer electronics
(refrigerator, stereo CD player). Identify the major components that define the assembly.
Define important relationships with *Equations* and Link Values.

## Exercise 6.3
## Create an OUTLET BOX

Design Situation. The current electrical box configuration consists of 4 parts:

- C-shaped plate

- Left flat plate

- Right flat plate

- Bracket

The flat plates are welded at each end of the C-shaped plate, Figure 6.E3. The bracket is then welded to the Right plate.

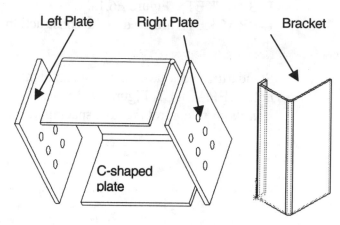

Figure 6.E3

The design challenge is to reduce four components into one flat sheet metal part that are formed into a 3D box.

Start with a Solid box. Create the BASE feature *symmetrical* about the *Right* plane.
- Add two Tabs.
- Center a mounting hole, Figure E6.3a.
- Create a mounting bracket on the right side, Figure E6.3b.
- Complete the OUTLET BOX. Add a keyway die cut. Add a dimple form, Figure E6.3c.

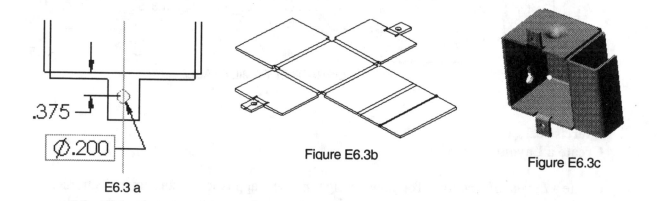

.375

Ø.200

E6.3 a

Figure E6.3b

Figure E6.3c

**Engineering Changer Order (ECO)**

| D&M — Engineering Change Order | ECO # _____ Page 1 of __ |
|---|---|

| | | Hardware | | Author |
| | | Software | | Date |
| Product Line | | Quality | | Authorized Mgr. |
| | | Tech Pubs | | Date |

Change Tested By

Reason for ECO( Describe the existing problem, symptom and impact on field)

| D&M Part No. | Rev From/To | Part Description | Description | Owner |
|---|---|---|---|---|
| | | | | |
| | | | | |
| | | | | |
| | | | | |
| | | | | |
| | | | | |
| | | | | |

| ECO Implementation/Class | | Departments | Approvals | Date |
|---|---|---|---|---|
| All in Field | | Engineering | | |
| All in Test | | Manufacturing | | |
| All in Assembly | | Technical Support | | |
| All in Stock | | Marketing | | |
| All on Order | | DOC Control | | |
| All Future | | | | |
| Material Disposition | | ECO Cost | | |
| Rework | | DO NOT WRITE BELOW THIS LINE (ECO BOARD ONLY) | | |
| Scrap | | Effective Date | | |
| Use as is | | Incorporated Date | | |
| None | | Board Approval | | |
| See Attached | | Board Date | | |

## SolidWorks Gold Partners (www.SolidWorks.com)

### Comprehensive Listing of All SolidWorks Gold Partner Products (June 2000)

**Analysis/FEM/FEA**

**COSMOS/Works** by Structural Research & Analysis Co
**FloWorks** by NIKA-USA
**visualNastran FEA for SolidWorks** by MSC Working
Knowledge

**Animation**

**SolidWorks Animator** by SolidWorks Corporation

**Cable and Harness Design**

**EMbassyWorks** by Linius Technologies

**Component Design/Libraries**

**SolidMech** by EMT Software, Inc.
**SolidParts** by SolidPartners, Inc
**Toolbox/SE** by CIMLOGIC

**Insp. Reverse Engineering**

**FeatureWorks** by SolidWorks Corporation
**RevWorks for SolidWorks** by Design Automation, Inc.

**Kinematics and Dynamics**

**Dynamic Designer/Motion** by Mechanical Dynamics Inc.

**Manufacturing**

**CAMWorks** by TekSoft CAD/CAM System, Inc.

**Mechanical Engineering**

**MechSoft-PROFI for SolidWorks** by MechSoft.com

**Mold Design and Analysis**

**Moldflow Plastics Advisers** by Moldflow Corporation
**MoldWorks** by R&B Ltd.
**Toolbox/MB** by CIMLOGIC

**Product Data Management**

**Activault** by SolidPartners, Inc
**DBWorks** by MechWorks
**eMatrix SolidWorks** by MatrixOne, Inc.
**PDM/Works** by DesignSource Technology, Inc
**SmarTeam-Works** by Smart Solutions
**SolidPDM** by Modultek Inc.
**WTC ProductCenter Integrator for SolidWorks** by
Workgroup Technology Corporation

## Services/Consulting

**Engineering Services** by MSC Technologies
**Engineering Services** by IMPACT Engineering Solutions
**MCAD Engineering Services** by Express CAD Engineering, Inc.

## Tolerance Analysis

**Sigmund ID** by Varatech

## Viewing, Rendering, and Collaboration

**ConceptWorks** by RealityWave Inc.
**PhotoWorks** by SolidWorks Corporation

Additional SolidWorks Solutions Partners are also listed on the SolidWorks Web site.

The SolidWorks web site (www.solidworks.com) and SolidWorks users groups provide additional information on using SolidWorks.